Mathematical Puzzles

STEPHEN AINLEY

G. BELL & SONS LTD.
LONDON

First published in 1977 by
G. Bell & Sons Ltd
37-39 Queen Elizabeth Street
London, SE1

ISBN   0 7135 1954 1

Filmset in Ireland by
Doyle Photosetting Ltd., Tullamore

Printed in Great Britain by
The Camelot Press Ltd Southampton

To M.G.

Contents

Introduction

I want to take this chance to express thanks to a number of people in and around this curious and addictive field of recreational mathematics: to the still lively shades of Lewis Carroll, Rouse Ball, Dudeney and Sam Loyd; to Eddington for his marvellous cricket problem; to Hardy and Littlewood and Coxeter for tucking so much pleasure into serious maths; and to two M.G.s included in my dedication for providing us with so regular and varied a diet. The diet we want—to speak for my own kind of addict—is not just tidy morsels from established recipes, tasty and welcome as those are, but pushings out into fresh areas and quiet encouragement to try our own hand at developing new seasonings and new dishes. And that is what the Mathematical Gazette and Martin Gardner consistently give us.

As for the puzzles in this book, I make one particular request to the reader: please don't take my solutions too much on trust. Like chess problems, puzzles—even those set by the eminent and excellent—can quite often be 'cooked'. I have checked what I say with some care, and I haven't consciously cheated, but I shall be surprised if I haven't made mistakes. And I really should like to hear from anyone who disagrees with an answer I have given, or can improve on it. So far as I am able, I shall be glad to answer letters of that kind.

Bon appétit!

(and, if you go on from a solution to think 'Ah, but what if . . .', bonne cuisine!)

STEPHEN AINLEY

A   Various

A1   Nik's primes

(1) 'What is a prime?' said Nik.

'A whole number greater than one,' I answered cautiously, 'whose only divisors are itself and unity.'

'Right,' he said, 'then find me three whole numbers in arithmetical progression whose product is a prime.'

Eventually, I did so.

(2) 'Now,' he said, 'find me a different 3-term arithmetical series whose product is that same prime. This time the terms need not be whole numbers: they can be fractions.'

(3) A good deal later, he said 'Finally, find me two different 3-term arithmetical series, each of which multiplies out to 11.'

Can you help me?

A2   Touching pennies

If $T(n)$ is the fewest number of pennies with which you can compose an arrangement with each penny lying flat on the table and each touching exactly n others, clearly $T(1) = 2$.

(1) What is $T(2)$?

(2) What is $T(3)$?

(3) What is $T(4)$?

If $S(n)$ is the fewest number when you may, in addition, put pennies on top of others,

(4) What is $S(3)$?

(5) What is $S(4)$?

A3   Converting tries

The goal-posts G1, G2 at Llantwizzock are 18 feet apart (See Fig. 1). Dai scored a try at T, x feet wide of the posts. He placed the ball for

1

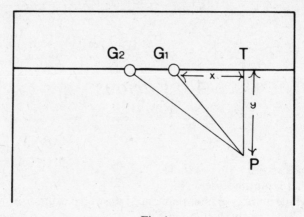

Fig. 1.

his conversion kick at P, a distance of y feet out from the goal-line, the angle GTP being, of course, 90°. When he missed the kick, although there was no wind to blow the ball off course, argument raged as to y. 'It should be equal to x,' said Fred; '$x+18$,' said Ned; '$x+9$,' said Ted.

On the assumptions—valid perhaps for Llantwizzock only—that the ball and the posts can be treated as having negligible dimensions and that P should be chosen to make the angle G_1PG_2 as large as possible, who was right?

(If x was 32 feet, for instance, Fred says y should be 32 feet, Ned that it should be 50 feet and Ted 41 feet.)

A4 Indivisible number-trios
If I want to find three different digits such that, in whatever way I arrange them into a 3-digit number, it is not divisible exactly by 3 or 5 or 7 or 11 or 13 or 17, you might suggest (1, 2, 4) or (1, 4, 9) or— what third set could you suggest?

A5 Balancing cards
We are trying here (See Fig. 2) to arrange n playing-cards so as to achieve the longest possible projection P(n) over the end of an oblong table. The cards may be arranged on top of one another in any way we please, but they must all be squarely lengthwise—long sides parallel to the long sides of the table. It is pretty clear that $P(1) = \frac{1}{2}$

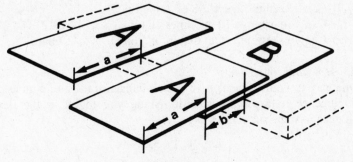

Fig. 2.

if the length of a card is 1. What is P(2)? With two cards arranged as shown $P(2)=a+b$. If you make a larger, b must be smaller and vice versa. Experiment and theory agree that $(a+b)$ is greatest if $a=\frac{1}{2}$, $b=\frac{1}{4}$, indicating $P(2)=\frac{3}{4}$.

(1) Is P(2) really $\frac{3}{4}$? Is there any other arrangement which gives as big—or a bigger—projection?
(2) What is P(3)?
(3) What is P(4)?

A6 Parity of a random sum

First choose two different whole numbers at random. Call them a and b, with b the bigger. Now choose two different numbers at random out of the range from a to b (both included) and add them. Is the resulting sum more likely to be odd or even?

A7 Chessboard configurations

Fig. 3 shows 6 pawns arranged on a 5×5 board, so that there are 4 lines of three (with each pawn in 2 of the lines).

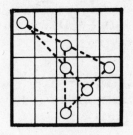

Fig. 3.

(1) Can you find other ways of doing that?
(2) Can you arrange 10 in 5 lines of 4 on a 7×7 board?
(3) Can you arrange 15 in 6 lines of 5 on an 11×11 board?

A8 How to be intelligent

One way to be intelligent is to pass intelligence tests, of course ...
Intelligence tests usually include asking you to fill in the next
number in some such series as

$$4, 5, 10, 19, 32, —$$

This irritates some, who rightly say that there is an infinite number of
legitimate 'next numbers'. But there is always a best answer if you
make the assumption that the given numbers are values derived by
putting $x = 0, 1, 2, 3$, etc. in a formula

$$f(x) = C_0 + C_1 x + C_2 x^2 + C_3 x^3 \ldots$$

in which the highest power of x is as low as possible. On that basis,
you can proceed by 'differences' as follows:

$$
\begin{array}{llll}
4 = D_0 \\
5 & 1 = D_1 \\
10 & 5 & 4 = D_2 \\
19 & 9 & 4 & 0 = D_3 \\
32 & 13 & 4 & 0 & 0 = D_4 \\
[49] & [17] & [4] & [0]
\end{array}
$$

Then

$$C_0 = D_0$$
$$C_1 = D_1 - \tfrac{1}{2}D_2 + \tfrac{1}{3}D_3 - \tfrac{1}{4}D_4 + \tfrac{1}{5}D_5 - \tfrac{1}{6}D^6 \ldots$$
$$C_2 = \tfrac{1}{2}D_2 - \tfrac{1}{2}D_3 + \tfrac{11}{24}D_4 - \tfrac{5}{12}D_5 + \tfrac{137}{360}D_6 \ldots$$
$$C_3 = \tfrac{1}{6}D_3 - \tfrac{1}{4}D_4 + \tfrac{7}{24}D_5 - \tfrac{5}{16}D_6 \ldots$$
$$C_4 = \tfrac{1}{24}D_4 - \tfrac{1}{12}D_5 + \tfrac{17}{144}D_6 \ldots$$
$$C_5 = \tfrac{1}{120}D_5 - \tfrac{1}{48}D_6 \ldots$$
$$C_6 = \tfrac{1}{720}D_6 \ldots$$

So, in our example, $C_0 = D_0 = 4$

$$C_1 = D_1 - \tfrac{1}{2}D_2 = 1 - 2 = -1,$$
$$C_2 = \tfrac{1}{2}D_2 = 2,$$

and the formula is $2x^2 - x + 4$: so the next number is $2.5^2 - 5 + 4 = 49$.

If you only want the next number, not the formula, you can simply write in the extra numbers shown in [ ] in the table.

(1) What is the next number in the following series?

$$1, 16, 81, 256, -$$

(2) If you join all the vertices of a *nearly* regular n-gon, how many intersection-points do you get, excluding the original vertices? (By 'nearly regular', I mean that no more than 2 lines cross at a single point.)

A9 Semi-numbers
(1) The number 2 is 'semi-1', because exactly half of the whole numbers up to 2 (i.e. the numbers 1 and 2) contain the digit 1, and half don't. 16 is also semi-1.

What is the largest semi-1 number?

(2) The smallest semi-2 number is clearly 2. What is the next smallest?

A10 How agy?
'Now,' begins Betty Lou, beefy abbess at Billowy Hill Abbey, 'How agy am I? A hint? I know I am as dotty at a ditty box as Abbot Benny is dotty at beer! Nor do I allow my boy Billy any accent at all! Now, how agy am I?'

I can't say why the abbess says 'agy' when she means 'old': but can you say how old she is—or, at least, how old she would say she is?

A11 Tiddly-wink all-sorts
I sell tiddly-winks in opaque packets, with N winks in each. My winks come in C different colours, and I market in equal quantities packets containing every possible combination of colours. For instance, if N were 2 and C were 3, I should sell bags containing XX, YY, ZZ, XY, XZ, YZ [X, Y, Z are the 3 colours], and you would be just as likely to get one type as another for your 1p.

Now suppose you buy a packet and, without looking, take out one random wink and find it is pink. Then you take out a second wink at random. Is it more likely to be pink or white? (I may add that I do sell white winks.) What can you say about the comparative likelihood of that second wink being pink or white? Can you, perhaps, give the comparative probability as a function of C and N?

A   Solutions

A1S Nik's primes

(1) $-3, -1, +1$: product 3.

(2) $1, \frac{3}{2}, 2$: product 3.

(3) $-4\frac{1}{2}, -1\frac{1}{3}, +1\frac{5}{6}$: product 11.

$\quad 1\frac{5}{6}, 2\frac{1}{4}, 2\frac{2}{3}$: product 11.

If the 3 numbers are $(a-d)$, a and $(a+d)$, we must solve

$$a(a^2 - d^2) = p(\text{prime}).$$

First, in integers, a must be ± 1. If $a = +1$, $p = 1 - d^2$: no solution. If $a = -1$, $p = (d+1)(d-1)$, yielding just the solution at (1).

In fractions, put $a = s/t$, $d = x/y$, with both fractions in lowest terms. We then arrive at $s^3 y^2 = t^2 (pty^2 + sx^2)$; so t^2 must divide y^2, i.e. $y = kt$. This leads to $sx^2 = k^2(s^3 - pt^3)$; so k^2 (which has no common factor with x^2) must divide s, i.e. $s = k^2 l$. Hence $pt^3 = l(k^6 l^2 - x^2)$; and l (which has no common factor with t^3) must divide p. So $l = 1$ or $l = p$. This gives two possible schemes for solutions:

(a) $pt^3 = (k^3 + x)(k^3 - x)$; $a = \dfrac{k^2}{t}$, $d = \dfrac{x}{kt}$;

(b) $t^3 = (k^3 p + x)(k^3 p - x)$; $a = \dfrac{k^2 p}{t}$, $d = \dfrac{x}{kt}$.

Of these, (a) yields a host of solutions, including those given at (3)—with $k2$, $x19$, and $k3$, $x5$ respectively—and (b) yields the solution at (2) and probably others as well.

A2S Touching pennies

(1) $T(2) = S(2) = 3$

(2) $T(3) = 16$

(3) $T(4)$ does not exist, I think

(4) $S(3) = 4$

(5) $S(4) = 6$

See Fig. 4, where dotted pennies are in an upper layer.

A3S Converting tries

Ted was nearest, but none was exactly right. The answer is that $y = \sqrt{x(x+18)}$. If x is 32 feet for instance, $y = 40$ feet.

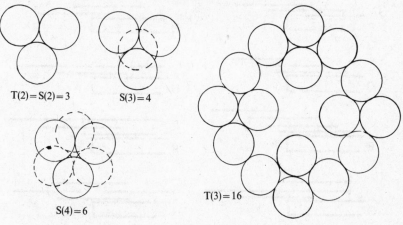

T(2) = S(2) = 3   S(3) = 4

S(4) = 6

T(3) = 16

Fig. 4.

The angle G_1PG_2 is $\sin^{-1}(18/(2x+18))$. So, for instance, to achieve a $30°$ angle, you must score 9 feet wide and place the ball $9\sqrt{3} = 15.59$ feet away from the goal-line.

A4S Indivisible number-trios
(2, 4, 8) is, I think, the only answer. Since each of its digits is double the corresponding digit in (1, 2, 4), an odd number that won't divide an arrangement of one clearly won't divide an arrangement of the other.

A5S Balancing cards
(1) Yes, $P(2) = \frac{3}{4}$. In Fig. 5, A shows the arrangement discussed in the question, but this time in profile. B is just as good.

(2) $P(3) = 1$ (as at D).

(3) $P(4) = \dfrac{15 - 4\sqrt{2}}{8}$, about 1.168 (as at I).

The method, illustrated by A, C and E in Fig. 5, of building from the top down, with successive projections of $\frac{1}{2}$, $\frac{1}{4}$, $\frac{1}{6}$..., is often quoted in discussions of this problem or the similar problem with dominoes or bricks instead of cards. And it—call it 'method S'—certainly is a neat way of getting an overall projection as long as you wish. But for 3 cards, D gives a projection of 1, against $\frac{11}{12}$ by method

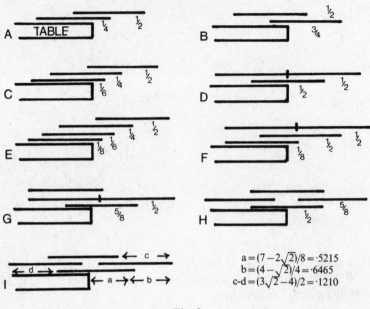

$$a = (7 - 2\sqrt{2})/8 = \cdot5215$$
$$b = (4 - \sqrt{2})/4 = \cdot6465$$
$$c\text{-}d = (3\sqrt{2} - 4)/2 = \cdot1210$$

Fig. 5.

S. For 4 cards, F, G and H show various ways of getting 9/8, compared with 25/24 in E: and I does even better, giving $(15 - 4\sqrt{2})/8$, or just over 7/6.

That is theory, of course, but as theory just as true for bricks or dominoes as for cards. In practice, D and F won't work for bricks or dominoes, but with cards, a tiny overlap of the top cards makes them work. And G, H and I are workable with all three materials.

What about larger values of N? I expect more detailed work would produce further improvements, but even if it did not we could always replace the top 4 elements of a method—S arrangement by 4 elements arranged as in I—or as in H, if the rusty statics underlying I have betrayed me. So method S never gives the maximum projection for more than 2 cards, bricks or dominoes, either in theory or in practice, I think.

A6S Parity of a random sum

Whatever a and b, the sum is more likely to be odd. It is not hard to work out the odds as follows:

	Number of choices making sum odd	Number of choices making sum even	Odds on an odd sum
a and b of opposite parity (i.e. one odd, one even)	$\dfrac{(b-a+1)^2}{4}$	$\dfrac{(b-a+1)(b-a-1)}{4}$	$\dfrac{b-a+1}{b-a-1}$
a and b of same parity (both even or both odd)	$\dfrac{(b-a)(b-a+2)}{4}$	$\dfrac{(b-a)^2}{4}$	$\dfrac{b-a+2}{b-a}$

If the range a to b is not too great, the odds are, perhaps surprisingly, quite significant for betting purposes.

A7S Chessboard configurations
Fig. 6 shows solutions.

A8S How to be intelligent
(1) On the principle of keeping the highest power of x as low as possible, the answer is 601: the formula being $10x^3 + 5x^2 + 10x + 1$. The answer the tester wants is probably 625, the formula being $(x+1)^4$. It is a matter of taste which is the better answer.

(2) $\dfrac{n^4 - 6n^3 + 11n^2 - 6n}{24} = \dfrac{n(n-1)(n-2)(n-3)}{24}$

Scribbling produces the following results (note how useful the easy-to-find zero values are):

```
n   f(n)
0   0
           0
1   0            0
           0            0
2   0            0           1
           0            1           0
3   0            1           1           0
           1            2           0           0
4   1            3           1           0
           4            3
5   5            6           1
           10           4
6   15          10
           20
7   35
```

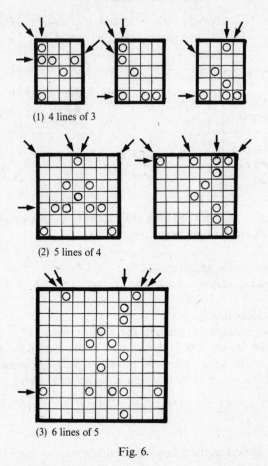

(1) 4 lines of 3

(2) 5 lines of 4

(3) 6 lines of 5

Fig. 6.

So $D_0 = D_1 = D_2 = D_3 = 0$, $D_4 = 1$, and the result follows.

Of course, you can in this case argue more simply that each set of 4 vertices corresponds to just one intersection, which gives the formula immediately, but the difference technique is an invaluable one for many 'how many?' problems.

A9S Semi-numbers
(1) 1,062,880
(2) 2914
I think the following list is complete:

Semi-1	Semi-2	Semi-3	Semi-4	Semi-6
2*	2	39,364*	472,390*	6,377,290*
16*	2,914*	41,288	630,226	
24	3,150	41,308	642,976	Semi-7
160*	3,152	41,558	671,782	7,440,172*
270	3,238	43,738	671,784	
272	3,398	44,686	4,251,526*	Semi-8
1,456*	26,242*	354,292*		8,503,054*
3,398	41,558	671,782		
3,418	42,280	671,784	Semi-5	Semi-9
3,420	44,686	673,594	590,488*	9,565,936*
3,422	236,194*	674,910	630,226	
13,120*	671,784	3,188,644*	656,098	
44,686	672,136		671,782	
118,096*	674,910		5,314,408*	
674,934	674,912			
1,062,880*	674,926			
	674,934			
	2,125,762*			

Those marked * are of the form (for 'semi-k') $2(k. 9^n - 1)$.

A10S How Agy?

'Forty' earns full marks. Betty Lou only uses words whose letters are in alphabetical order—that is why she says 'agy' for 'old'—and 40 is the only number she can utter in English. But I would award full marks plus a bonus to the answer 'Almost forty'.

A11S Tiddly-wink all-sorts

The second wink is twice as likely to be pink as white. The chance of its being pink is $2/(C+1)$: the chance of its being white is $1/(C+1)$. These chances are independent of N.

You can satisfy yourself that this is so for a particular value of C and N, as follows. Take N3, C2. Now calculate as in the following table:

(1) Assortment	PPP	PPW	PWW	WWW	*Total*
(2) Chance that the first pink wink came from this assortment	$\dfrac{3}{6}$	$\dfrac{2}{6}$	$\dfrac{1}{6}$	$\dfrac{0}{6}$	
(3) Chance that a second wink from this packet will be P	$\dfrac{2}{2}$	$\dfrac{1}{2}$	$\dfrac{0}{2}$	$\dfrac{0}{2}$	
(4) Chance that a second wink will be W	$\dfrac{0}{2}$	$\dfrac{1}{2}$	$\dfrac{2}{2}$	$\dfrac{2}{2}$	
(5) Chance of thus getting P/P $(=(2)\times(3))$	$\dfrac{2.3}{12}$	$\dfrac{1.2}{12}$	$\dfrac{0.1}{12}$	$\dfrac{0.0}{12}$	$\dfrac{8}{12}$
(6) Chance of thus getting P/W $(=(2)\times(4))$	$\dfrac{0.3}{12}$	$\dfrac{1.2}{12}$	$\dfrac{2.1}{12}$	$\dfrac{2.0}{12}$	$\dfrac{4}{12}$

and I think that points the way to a formal proof.

B Practical Geometry

B1 Alf and Buridan

In Fig. 7, A is Alf, the fly, and B is Buridan the spider, at diagonally opposite corners of the floor of the room. Alf is asleep. Buridan wants to walk from B to A by the shortest path on the walls and/or ceiling; he avoids the floor, lest he be trodden on. He finds to his surprise that there are 5 distinct equal shortest paths from B to A.

The room is 8 feet high. What are its length and breadth?

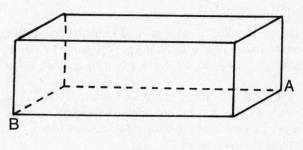

Fig. 7.

B2 The biggest field

'I will give you a present,' said Uncle Jack, 'of a field with 4 straight sides, whose lengths, in whatever order you chose, are to be 1, 2, 8 and 9 furlongs.' Assuming I want the largest possible field,

  (1) In what order should the sides be?
  (2) What shape should the field be?
  (3) How many square furlongs shall I get? And can you tell me
  (4) What other set of 4 whole-number side-lengths would give me the same area?

13

B3 Rail networks
(1) A, B, C, D, E, and F are 6 towns located at the corners of a double square. Fig. 8 shows a plan for a rail network linking all 6 towns. As you see, it is exactly 500 miles long. Can you design a shorter network? You can if you wish have junctions away from towns. What is the shortest network which will link:
  (2) Four points at the vertices of a square?
  (3) Five at the vertices of a regular pentagon?
  (4) Six at the vertices of a regular hexagon?

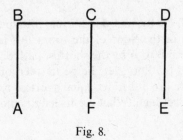

Fig. 8.

B4 Buridan's third bun
Buridan has to collect three buns and return to his starting-point. He wants to make the shortest journey. The ground is flat. From his starting point, bun A lies 69 feet due North, and bun B lies 92 feet due East.

   In what order is Buridan to collect the buns? ABC, ACB, BAC, BCA, CAB or CBA? It turns out that he has exactly the same length of journey, whichever of the 6 orders he adopts.

   How long *is* his journey?

B5 Table behind sofa
The back of my sofa is 6 feet long. I want to push it across the corner of the room, touching both walls, and put behind it the largest possible square-topped table.

   How should I arrange the sofa to accommodate the largest possible table (T) i.e. how long will OA and OB be? Which of the two arrangements in Fig. 9 will hold the larger table?
And what will the area of the table-top be?

B6 Simon's box and drawer
'This box and this drawer,' said Simon to himself, 'are both rect-

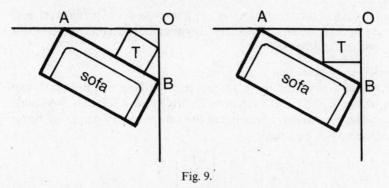

Fig. 9.

angular, and both are an exact number of inches wide and long. But how am I to get the box into the drawer? Height is no problem; the box is a tiny bit shallower. But the box is longer than both the width and the length of the drawer ... '

After a little thought, however, he did just manage to fit the box in, though it was a very tight fit.

What was the smallest size the box could have been?

B7 Square in triangle
In Fig. 10, ABC is an acute-angled triangle, with the largest square which will fit into it, based on side *a*, drawn in. In such a triangle, the side of the largest square based on side k is $2kT/(k^2 + 2T)$, in which T is the area of the triangle.

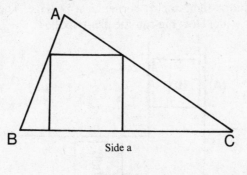

Fig. 10.

If I asked you to draw a triangle of such proportions that the size of the largest square that can be fitted into it is the same, regardless

of which of the three sides it is based on, you would perhaps propose
an equilateral triangle and rightly enough. Is there any other shape
you could propose?

B8   Five-point tour

I want the shortest route, starting and finishing at A in Fig. 11, and
visiting B, C, D, and X en route. Clearly I have $4 \times 3 \times 2 = 24$ possible
routes to consider. I find that of these 6 are equally good, and better
than any of the others.

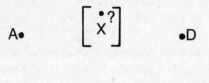

Fig. 11.

A, B, C and D are, as you see, at the corners of a rectangle. AB is
1 mile. AD is 2 miles.

Where is X? And how long is my trip?

B9   Fish-tanks round corners

Fig. 12 shows my square fish-tank before (A) and after (B) being
moved round a right-angled corner in a passage. The passage is
exactly 4′3″ wide. How big can the fish-tank be?

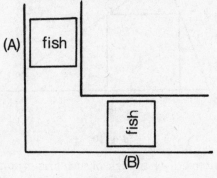

Fig. 12.

If I didn't mind which way the tank faced after the move, clearly it could be very nearly 4′3″ square itself. It could move directly 'south' into the corner and then directly 'east' to position B. But supposing that

(1) It is to end up as shown, having rotated a ¼-turn *anti-clockwise*
(2) It is to end up having rotated a ¼-turn *clockwise*

how long can the tank's side be? And finally

(3) What is the size and shape of the largest fish-tank (rectangular, but not necessarily square) which could be moved round the corner? Answers correct to the nearest inch will do. No tilting is allowed.

B10   Fire cover at Drayton Green

Mr Herne Hill, the Fire Commissioner of Drayton Green, which is 100 miles square, took over 5 fire-stations, placed as shown in Fig. 13, one at each corner and one in the middle. Proceeding as the crow

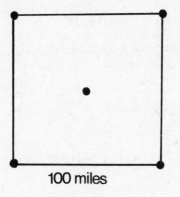

100 miles

Fig. 13.

flies, his men had thus 50 miles at most to travel to reach a fire anywhere in the Green. He rearranged them, so as to reduce the maximum distance to reach any fire to . . . what distance? And how did he arrange them?

B   Solutions

B1S   Alf and Buridan

24 feet long and 16 feet wide.

If you open up the room, hingeing the walls up flat with the ceiling, you get Fig. 14. The 5 paths can only be the two routes from B to A along the bottom of the walls, *just* clear of the floor, each $(l+b)$ feet long, and the 3 routes dotted across walls and ceiling.

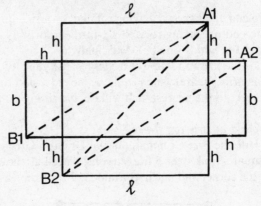

Fig. 14.

So $l+b=\sqrt{(l+h)^2+(b+h)^2}=\sqrt{l^2+(b+2h)^2}$. Squaring, we find that $lb=lh+bh+h^2=2bh+2h^2$

∴ $lh=h^2+bh$

∴ $l=b+h$.

Also $b(b+h)=2h(b+h)$

∴ $b=2h$

∴ $l=2h+h=3h$.

Since $h=8$ feet, $b=16$ feet and $l=24$ feet.

B2S   *The biggest field*

(1) It doesn't matter.

(2) Cyclic i.e. all 4 vertices lying on a circle.

(3) 12.

(4) 1, 3, 6 and 8.

As to side-order, Fig. 15 shows that whatever area can be achieved with the 'best' side-order *abcd* can also be achieved with order *abdc*: and similarly with order *adbc*. So order is immaterial.

As to shape, trigonometry tells us that, if the area is S, the semi-perimeter s, and the angles X and Y as shown,

$$S^2=(s-a)(s-b)(s-c)(s-d)-abcd\cos^2\left(\frac{X+Y}{2}\right).$$

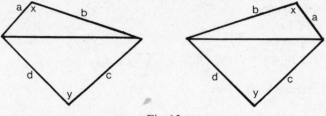

Fig. 15.

This is clearly biggest, for fixed a, b, c and d, when $\cos((X+Y)/2)=0$, i.e. when $X+Y=180°$, i.e. when the quadrilateral is cyclic. Area then comes from that formula, and is $\sqrt{9.8.2.1}=12$. Finally, sides 1, 3, 6 and 8 yield area $\sqrt{8.6.3.1}=12$ also.

B3S   Rail networks
Fig. 16 shows the best patterns.
(1) About 462.5 miles. P is $((9+\sqrt{3})/39)\times 100$ miles east of AB, and $((33-5\sqrt{3})/39)\times 100$ miles north of AF. The angle ABP is $36°12'$.

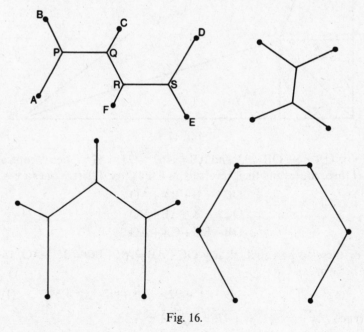

Fig. 16.

(2) 273·2 miles
(3) 389·1 miles
(4) 500 miles

At all the junctions, each angle is 120°.

B4S   Buridan's third bun
288 feet.
In Fig. 17, if O is the starting-point, and A, B and C the buns, we

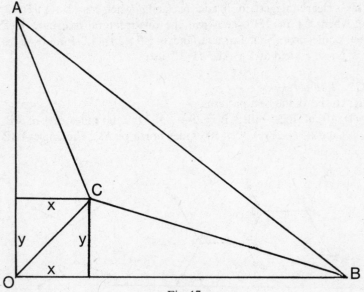

Fig. 17.

know OA = 69, OB = 92, and AB = $\sqrt{69^2 + 92^2} = 115$. The six routes are three routes and their reversals, and the three different ones are—

$$OC + CB + BA + AO$$
$$OC + CA + AB + BO$$
$$OB + BC + CA + AO.$$

For these to be equal, clearly OC + AB = AC + BO = BC + AO. In other words,

$$\sqrt{x^2 + y^2} + 115 = \sqrt{x^2 + (69-y)^2} + 92 = \sqrt{y^2 + (92-x)^2} + 69 \qquad (1)$$

Hence $\sqrt{x^2 + y^2 - 138y + 4761} = \sqrt{x^2 + y^2} + 23.$

Squaring, we find, after simplifying, $\sqrt{x^2 + y^2} = 92 - 3y$.
Hence also

$$\sqrt{x^2 + y^2 - 184x + 8464} = \sqrt{x^2 + y^2} + 46; \quad \text{and} \quad \sqrt{x^2 + y^2} = 69 - 2x.$$

So $x = (3y - 23)/2$, and hence, from (1), $y = 21$, and $x = 20$.
So $OC = \sqrt{21^2 + 20^2} = 29$, and the journey $= 29 + 75 + 115 + 69$
$$= 29 + 52 + 115 + 92$$
$$= 92 + 75 + 52 + 69$$
$$= 288.$$

B5S   Table behind sofa

The arrangement with the table squarely in the corner (on the right in Fig. 9) with OA and OB equal (each $3\sqrt{2}$ feet), permits a square table-top of $4\frac{1}{2}$ square feet in area. With the other arrangement, the table-top area is at most 4 square feet.

In any right-angled triangle, with sides a and b enclosing the right angle, the side of the largest square which can be put in the triangle is not difficult to calculate as

$$x(A) = \frac{ab\sqrt{a^2 + b^2}}{a^2 + ab + b^2} \quad \text{with a square-side along the hypotenuse}$$

$$x(B) = \frac{ab}{a+b} \quad \text{with square-sides along sides } a \text{ and } b$$

If you square these expressions, you will see that $x(B)$ is always the larger. And, for a given hypotenuse, $(a^2 + b^2)$ is constant, and $x(B) = ab/(a+b)$ is greatest when $a = b$. Hence the answer.
Note. If we allowed a rectangular table-top, the problem would be less interesting. It would still be right to make OA and OB equal, so as to maximize the triangle behind the sofa. But either arrangement would then permit a table-top of $4\frac{1}{2}$ square feet: double-square in one case, square in the other.

B6S   Simon's box and drawer

A 20 in. × 5 in. box (in a 19 in. × 16 in. drawer): see Fig. 18.
The box must be placed cornerwise, surrounded by two pairs of similar right-angled triangles. The sides of an integral right-angled triangle are multiples of $(a^2 - b^2)$, $2ab$ and $(a^2 + b^2)$, with a and b coprime and of opposite parity. So we can write in dimensions as in

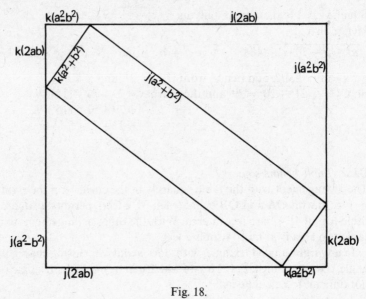

Fig. 18.

the figure. And we can assume $j > k$ as well as $a > b$. For the box to be longer than either dimension of the drawer,

    (1) $j(a^2 + b^2) > k(a^2 - b^2) + 2jab$; i.e., $j(a-b)^2 > k(a^2 - b^2)$,

and

    (2) $j(a^2 + b^2) > j(a^2 - b^2) + 2kab$; i.e., $2jb^2 > 2kab$.

So j/k must be greater than $(a+b)/(a-b)$ and greater than a/b.

    A little scribbling shows that we get the smallest box with $a = 2$, $b = 1$, in which case j/k must be greater than 3. So $j = 4$, $k = 1$ leads to the smallest box, with dimensions as given above.

B7S   Square in triangle

Yes: an obtuse-angled triangle with sides in the proportion $1 : 1 : 1{\cdot}5514$. The angle at C is about $101°45'$ See Fig. 19.

    Clearly, there is no other solution without turning to obtuse-angled triangles. In these, the square based on the longest side (c) has a side of $2cT/(c^2 + 2T)$, as with acute-angled triangles. But for squares based on the other sides we need a different formula: $x(a) = 4aT/(4T + a^2 - b^2 + c^2)$. So our only hope of an answer is to put $a = b$, and solve $4aT/(4T + c^2) = 2cT/(c^2 + 2T)$. With $a = b$, $T = (c/4)\sqrt{4a^2 - c^2}$, so

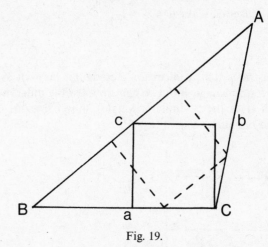

Fig. 19.

this reduces to $c^2(2a-c)=(c-a)^2(2a+c)$. Putting $a=1$, we have
$2c^3-2c^2-3c+2=0$,
whence $c=1\cdot5514$, approximately.

B8S   Five-point tour
X is $\frac{1}{2}$ mile 'north' of BC and $(1-(\sqrt{3}/3))$ of a mile (about $\cdot423$) 'east' of AB. See Fig. 20. My trip is thus $(4+(4\sqrt{3}/3))$, or about $6\cdot309$, miles long, whether I go AXBCDA, ABXCDA, ABCDXA, or the reverse of any of these.

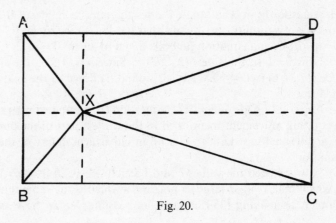

Fig. 20.

B9S Fish-tanks round corners
(1) 4 feet
(2) 4 feet
(3) 6 feet × 3 feet
Fig. 21 illustrates the problem for rectangular (as well as square) tanks, and for a passage with two arms of possibly different widths. Consideration of the 5 similar triangles shows that (a) $bl + pq = px + qy$, and of course (b) $p^2 + q^2 = l^2$.

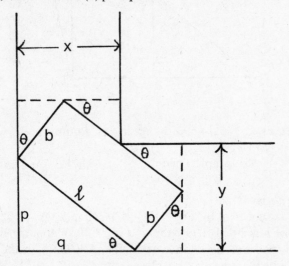

Fig. 21

The question is at what angle θ does the tank fit tightest? If (as here) $x = y$, it is intuitively obvious that $\theta = 45°$ is the critical point, so $p = q = l/\sqrt{2}$, and equation (a) boils down to (c) $2b + l = 2\sqrt{2}.x$.

(1) Now $b = l$, so the side $= (2\sqrt{2}/3).x$. Since $x = 4\frac{1}{4}$ feet, the side can be $(17\sqrt{2}/6)$ feet $= 4·006$ feet, or, rounding down to the nearest inch, 4 feet.

(2) The rotating movement as in question (1), combined with the non-rotating movement mentioned in the earliest part of the question, enables a 4-foot tank to end up in either arm in any of the 4 orientations.

(3) Now we need the values b and l which satisfy $2b + l = 17/\sqrt{2}$, and which make (bl) as large as possible — which is the same thing, obviously, as making ($2bl$) as large as possible. So we have two

numbers, $2b$ and l, whose sum is constant and whose product is to be a maximum. The way to do this is to make the two numbers equal. So $2b=l=17/2\sqrt{2}$.

So $l=6{\cdot}009$ feet, or 6 feet, to the nearest inch and $b=3$ feet, similarly.

Notes (1) tackling similar problems with x and y different involves more complicated difficulties; we have I think to differentiate (with respect to p) something like:

$$4p^4 - 4p^3 y - (4l^2 - x^2 - y^2)p^2 + 2l^2 yp + l^2(l^2 - x^2).$$

(2) If we put $b=0$ in equation (c), getting $l=2\sqrt{2}.x$, we get the longest 'plank', of negligible thickness, and rigid, which can be moved round the corner. For a passage 4 ft 3 in. wide, it is 12·019 feet.

B10S   Fire cover at Drayton Green
To just under 32·65 miles. With the layout in Fig. 22, the distances D are $\sqrt{1066}=32{\cdot}65$ miles, and P is only 32.28.

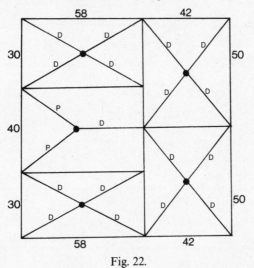

Fig. 22.

If instead of 42 we put 41·9, 58 becomes 58·1, 30 becomes 29·665, and so on, with D = 32·6175 and P = 32·524. So we could bring D down a little bit lower, but only a very little. A theoretical approach seems to lead to a quintic equation, which is insoluble to me (and to Mr Hill).

C Chess Pieces

C1 Black and white queens

Fig. 23 shows one attempt at putting as many White Queens (circles) as possible on a 5 × 5 board, so as to leave room for at least as many Black Queens (crosses), with no Queen attacked by a Queen of the other colour.

(1) Can I do better than that? (The pattern in Fig. 23 gives 3/4. I should regard 3/5 as better, and 3/6 as better still, but 4/4 as better than either.)

(2) What is the best solution for larger square boards, say 8 × 8 and 9 × 9?

(3) And what about an 'infinite' board? Mark out zones of two colours (say, red and blue) in a square, so that no blue point is in line vertically, horizontally, or diagonally with a red point. How large can you make the smaller zone?

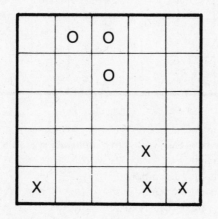

Fig. 23.

C2 Many-coloured queens

How many Queens of more than 2 colours can be placed on an ordinary 8 × 8 chessboard, with no Queen attacking a Queen of another colour?

We are trying, *first*, to place as many as possible of the colour of which there are fewest; *second*, as many as possible of the next fewest colour, and so on. For instance, figure 24 shows an attempt with 3 colours (0, 1 and 2). It scores 3/3/4. 3/3/5 would be better. 3/4/4 would be better still. 4/4/4 would be better again.

What is the best you can achieve for 3, 4, 5, 6, 7, 8 Queens on an 8 × 8 board?

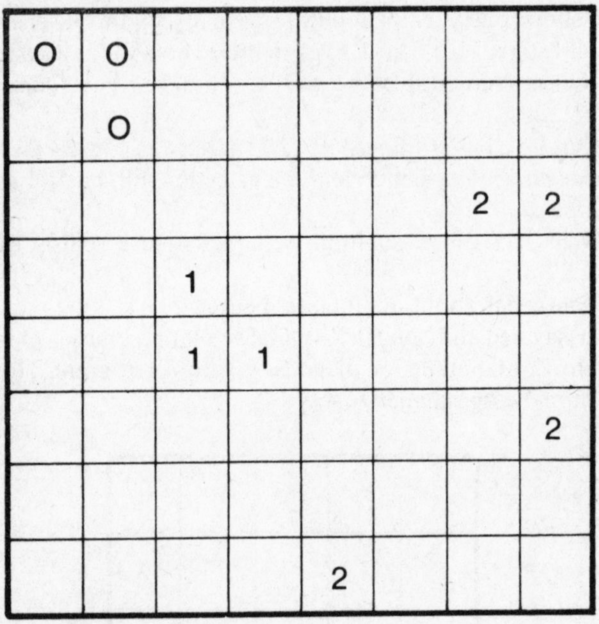

O	O						
	O						
						2	2
			1				
			1	1			
							2
				2			

Fig. 24.

C3 Efficient queens

How few squares can you leave unattacked, after placing with maximum 'efficiency'

(1) 2 Queens on a 5 × 5 board?
(2) 2 Queens on a 6 × 6 board?
(3) 3 Queens on a 7 × 7 board?
(4) 4 Queens on a 8 × 8 board?

C4   Inefficient queens

(1) First, how many queens can be placed on an 8 × 8 board without attacking all the unoccupied squares?

(2) How many on a 101 × 101 board?

(3) Now, how many squares can you leave unattacked after placing 8 queens as 'inefficiently' as possible on an 8 × 8 board?

(4) How many after placing 13 queens on a 13 × 13 board?

(5) How many after placing 14 queens on a 14 × 14 board?

C5   Few pacific queens

One queen, placed in the central square, will attack all unoccupied squares on a 3 × 3 board: Fig. 25 makes this obvious.

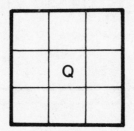

Fig. 25.

What is the smallest number of 'pacific' queens which can be placed so as to attack all the unoccupied squares (but not to attack each other), on larger square boards, from 4 × 4 up to 11 × 11?

Can you achieve a smaller number in any of these cases if the queens are not necessarily pacific?

C6   Knights, pacific and protected

What is the smallest number of knights which can be placed on a chessboard so that all the unoccupied squares are attacked, if

(1) All Knights are pacific (i.e. none attacks another)?

(2) All Knights are protected (i.e. every occupied square is attacked)?

(3) Knights may be either pacific or protected as you please?

Consider these 3 problems both for an ordinary 8 × 8 chessboard and also, if zeal permits, for a 6 × 6 board.

C7   Many pacific super-queens

A Superqueen moves like a queen *or* a knight. So it is not easy to place n Superqueens on a $n \times n$ board in such a way that none attacks another. Fig. 26 shows the only way, for instance, to put as many as 4 on a 5×5 board.

What is the smallest $n \times n$ board on which you *can* place n Superqueens 'pacifically'?

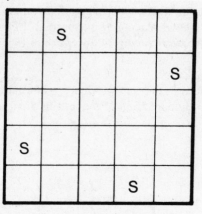

Fig. 26.

C8   *Bracers, many and few*

Bracers are powerful chess pieces, but only when there are 2 or more of them on the board. Any two bracers attack all squares whose centres are exactly on a straight line drawn through the centres of the bracers' squares. On the 9×8 board shown in Fig. 27 the squares x are attacked by the bracers B. You may never have 3 bracers in line.

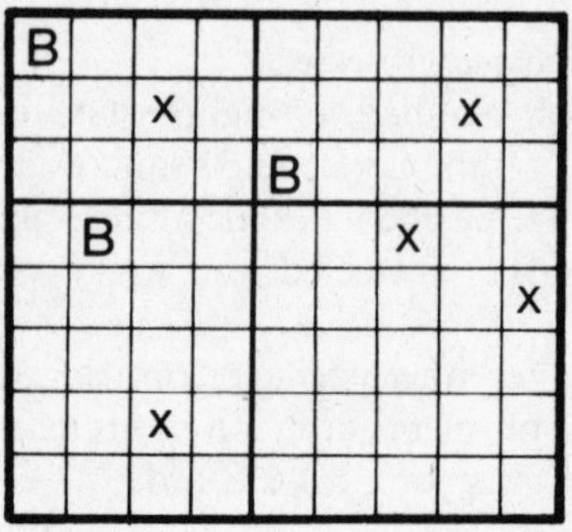

Fig. 27.

(1) How many bracers can you place on an 8 × 8 board? And how big a board can you completely command (i.e. attack all the unoccupied squares of) with the following?

(2) 6 bracers
(3) 8 bracers
(4) 10 bracers
(5) 12 bracers

(In deciding 'how big', the shorter dimension is the more important; 8 × 8 is 'bigger' than 7 × 12, but 8 × 9 is 'bigger' still.)

C Solutions

C1S Black and white queens

(1) Yes, you can get 4/4 on a 5 × 5 board. Figure 28(C) shows one way.

(2) You can get 9/10 on 8 × 8, and 12/12 on 9 × 9. Figure 28(D and E) shows one way of doing each.

(3) You can get $\frac{7}{48}$ red and $\frac{7}{48}$ blue on an 'infinite' board, as at A in figure 28. This is distinctly better than the sizes achieved with a single red blob in the corner, as at B, which give areas of $(2 - \sqrt{3})/2 = \cdot 134$.

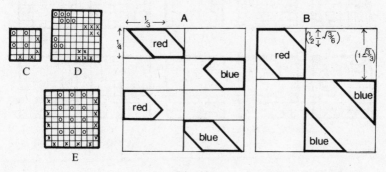

Fig. 28.

For most sizes of board, the '4-blob' approach exemplified at A and D gives better results than the 'sprinkle' approach of C and E. And actual 4-blob results correspond very closely indeed to the 'infinite' figure of $\frac{7}{48}$.

The following table lists my best results for 4-blob on an $n \times n$ board:

n	3	4	5	6	7	8	9	10	11	12	13	14	15	16
Queens	—	$\frac{2}{3}$	$\frac{3}{5}$	$\frac{5}{6}$	$\frac{7}{7}$	$\frac{9}{10}$	$\frac{11}{13}$	$\frac{14}{15}$	$\frac{17}{18}$	$\frac{21}{21}$	$\frac{24}{25}$	$\frac{28}{29}$	$\frac{32}{33}$	$\frac{37}{37}$
n	17	18	19	20	21	22	23	24	25	26	27	28	29	30
Queens	$\frac{42}{42}$	$\frac{47}{47}$	$\frac{52}{53}$	$\frac{58}{58}$	$\frac{64}{64}$	$\frac{70}{71}$	$\frac{77}{77}$	$\frac{84}{84}$	$\frac{91}{91}$	$\frac{98}{99}$	$\frac{105}{107}$	$\frac{114}{114}$	$\frac{122}{123}$	$\frac{131}{131}$

'Sprinkle' is better for $n=5$ or 9, and as good for $n=4$ or 6 or 13.

C2S   Many-coloured queens

Fig. 26 shows the best solutions I have found. That for 8 colours is not unique, certainly, and the others probably aren't. It is not hard to show, however, that those for 6, 7 and 8 colours cannot be improved on.

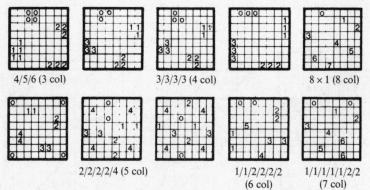

4/5/6 (3 col)   3/3/3/3 (4 col)   8 × 1 (8 col)

2/2/2/2/4 (5 col)   1/1/2/2/2   1/1/1/1/2/2
                            (6 col)   (7 col)

Fig. 29.

C3S   Efficient queens

(1) 2 unattacked by 2 queens on 5×5
(2) 6 unattacked by 2 queens on 6×6
(3) 4 unattacked by 3 queens on 7×7
(4) 2 unattacked by 4 queens on 8×8

Possible solutions are shown in Fig. 30.

Fig. 30.

C4S   Inefficient queens

(1) 42

(2) 9,900

(3) 11

(4) 44

(5) 56

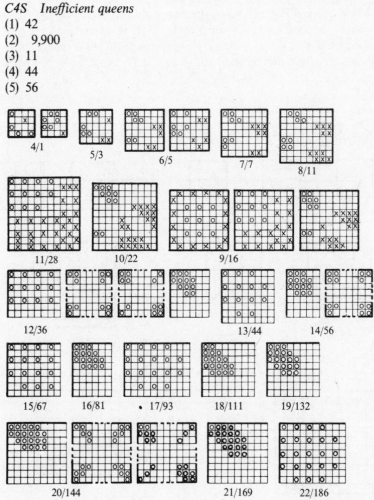

Fig. 31.

If questions (1) and (2) had been 'How many squares can be left unattacked by one queen placed as inefficiently as possible?', the answer would fairly clearly be got by putting the queen on any border square, where she attacks only $3(n-1)$ other squares, and the answer would therefore be $(n-1)(n-2)$. And this is the answer to these questions as put.

Questions (3), (4) and (5) are similar to the questions in Black and White queens. Fig. 31 shows the best solutions I have found for n queens on an $n \times n$ board for n up to 22. It is interesting that *here* 'four-blob' solutions, from $n = 10$ onwards, fall further and further behind 'scatter' and 'corner' solutions, which continue to run neck and neck.

C5S   Few pacific queens

The answers are shown in Fig. 32. Only when $n = 4$ or 6 is an improved un-pacific solution possible.

4 × 4 with 3Qs, or 2 (not pacific)

5 × 5 with 3 Qs

6 × 6 or 7 × 7
with 4 Qs

6 × 6 with 3
(not pacific)

8 × 8 or 9 × 9
with 5 Qs

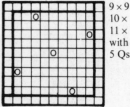

9 × 9
10 × 10
11 × 11
with
5 Qs

9 × 9
with
5 Qs
(use 0
and
4 × 1/2/3/4/5/6)

Fig. 32.

C6S   Knights, pacific and protected
Figure 33 shows the answers as follows:
  On an 8 × 8 board   15 Pacific, 14 Protected, 12 mixed.
  On a 6 × 6 board   8 in each case.

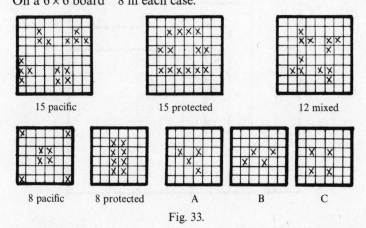

15 pacific   15 protected   12 mixed

8 pacific   8 protected   A   B   C

Fig. 33.

Only, I think, on the 8 × 8 board is it possible to improve on both
the pacific and the protected solution.

In the 'protected' problem, but not of course in the others, it is
possible to consider the black and white squares separately. For a
6 × 6 board, for instance, we get patterns A, B and C for one colour,
and we can combine these in various ways. In general, there are
several solutions for each 'protected' case.

The best results I have so far found for boards of $n \times n$ up to
10 × 10 are:

n	3	4	5	6	7	8	9	10
Pacific knights	–	4	5	8	13,	15	14	16
Protected knights	–	6	7	8	10	14	18	22

C7S   Many pacific superqueens
10 on a 10 × 10 board, as shown, seems to be the smallest case. The
bottom right-hand corner of figure 34 shows one way (not the only
one) to place 4 in 6^2, 5 in 7^2, 6 in 8^2, and 8 in 9^2, which I believe to be
the most achievable in each case.

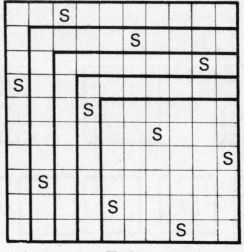

Fig. 34.

C8S   Bracers, many and few
Figure 35 shows the best answers I have found. That given for (1)
is one of four symmetrical solutions.

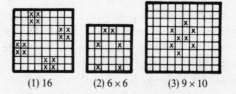

(1) 16   (2) 6 × 6   (3) 9 × 10

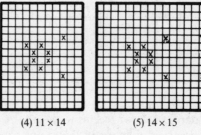

(4) 11 × 14   (5) 14 × 15

Fig. 35.

D   Square Routes

D1   Mini-rook

An old puzzle which shows the importance of old-fashioned school-masters' instructions to *read the question carefully* asks you to trace a path for a rook, which moves one square at a time, starting in one corner square of an ordinary chess-board and ending in the dia-metrically opposite corner square, so that it enters every square once and once only en route.

This is a similar problem on a mini-board of 16 squares (Fig. 36).

1	2	3	4
5	6	7	8
9	10	11	12
13	14	15	16

Fig. 36.

A rook starts in square 1 and moves one square at a time till it's in square 16, entering every square just once on the way. How many different paths can it take?

D2   Awkward spud squares

Nine potatoes are laid out in a square 3 × 3 pattern, one yard between

37

rows. I must pick them up in turn, in the order you specify, and then
return to the first. I must travel all the time along the grid lines. If you
specify the order shown on the left in Fig. 37, I shall only have to

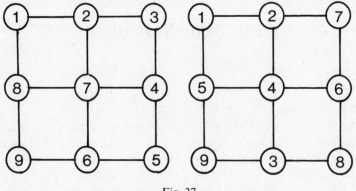

Fig. 37

travel 10 yards. But if you specify the more 'awkward' order shown
on the right, I shall have to travel 14 yards.

If the route you specify must always give me a straight north-south
or east-west route for each leg of my route, what is the most awkward
order you can specify for a square array of
   (1) 16 potatoes?
   (2) 64 potatoes?
You can, of course, think of these problems as finding the longest
re-entrant route for a rook on a n^2 chessboard.

   (3) What is the most awkward order you can specify for a *bishop's*
re-entrant route covering all the white squares on an 8×8 chessboard?

D3   Very awkward spud squares
Fig. 38 shows $3 \times 3 = 9$ potatoes laid out in a square grid. I must pick
them up in turn, in the order you specify, and then return to no. 1.
I can move along the grid-lines only. If you specify the order shown,
I have to travel $2+2+3+2+2+2+3+2+2 = 20$ yards. What is
the longest journey you can make me have in picking up
   (1) $3 \times 3 = 9$ potatoes?
   (2) $4 \times 4 = 16$ potatoes?
   (3) $8 \times 8 = 64$ potatoes?

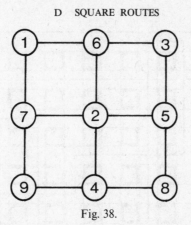

Fig. 38.

D4   Rooks' tours
In Fig. 39, the first diagram shows a rook's tour between A and B, the

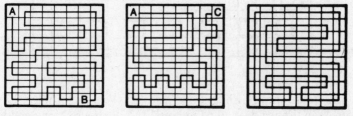

Fig. 39

two squares on an 8 × 8 chessboard which are farthest apart, visiting
every other square just once. (Such a tour between the two squares at
extreme ends of a main diagonal is not possible on a board with an
even number of squares.) The tour has 30 'legs'. The second diagram
shows a 31-leg tour between A and C. The third diagram shows a
20-leg re-entrant tour.

(1) What is the smallest number of legs with which you can achieve
each of the 3 kinds of tour?

(2) What is the largest number of legs with which you can do so?

D5   Shorter for Clancy
Clancy the policeman, according to Sam Loyd, has to patrol the 49
houses, starting and finishing at the arrowed point (see Fig. 40). He
has to pass an odd number of houses along any street or avenue
before he makes a turn, and he cannot walk twice along any part of
his route.

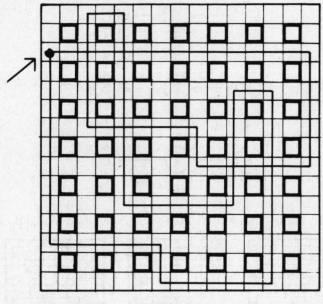

Fig. 40.

The route shown, which Loyd gives, answers the problem Loyd sets, 'Can Clancy pass all 49 houses?' It is 50 'houses' long and has 15 turns.

Can you help Clancy even more, and find him a shorter route, and the shortest route with the fewest turns?

D6   Square cascades
The first two pictures in Fig. 41 show the only two ways of fitting

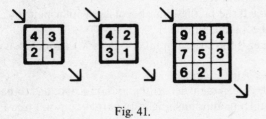

Fig. 41.

together four square tiles of different heights so that no tile has a higher tile to its east or its south. So the water from a fountain in square '4' will cascade prettily down over all the edges from the top left to the bottom right corner.

(1) The third picture shows one of the (*how many?*) ways to 'cascade' a 3×3 square.

(2) How many ways are there of 'cascading' a 4×4 square?

D   Solutions

D1S   Mini-rook
There are 8 paths.

The rook must start by moving to square 2 (or square 5) and then immediately back into square 1; otherwise it can't *enter* square 1 on the way. It is easy then to see that there are only 8 paths:

$$2, 1, 5, 6, 10, 9, 13, 14, 15, 11, 7, 3, 4, 8, 12, 16$$
$$2, 1, 5, 9, 13, 14, 10, 6, 7, 3, 4, 8, 12, 11, 15, 16$$
$$2, 1, 5, 9, 13, 14, 15, 11, 10, 6, 7, 3, 4, 8, 12, 16$$
$$2, 1, 5, 6, 7, 3, 4, 8, 12, 11, 10, 9, 13, 14, 15, 16$$

and their 4 reflections beginning $5, 1, 2 \ldots$

D2S   Awkward spud squares
For the 'rook' problem, the method exemplified in (1) and (2) in Fig. 42 will always produce a route of length $1 + n(2n^2 - 5)/3$ when

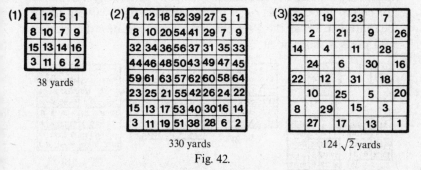

(1)   38 yards

(2)   330 yards

(3) $124\sqrt{2}$ yards

Fig. 42.

n is odd, and of $2 + n(2n^2 - 5)/3$ when n is even. The larger n is, the more different patterns are possible, but this type is particularly easy to write out. You start with 1, 2, 3, and 4 in successive corners, and then put each number as far as possible from its predecessor. If you have a choice between 2 legs of equal length, choose the one in the same line as the previous leg. The only complications are

(a) When you come to fill in the last row/column, insert number n^2 immediately after number $(n^2 - n + 1)$, and then go on with $(n^2 - n + 2)$ and so on:

(b) You may finally have to switch numbers 1 and 3 so as to bring number 1 into line with number n^2.

The 'bishop' problem seems trickier. I will do no more than give my own best results, as follows:

n	3	4	5	6	7	8	9	10
Length ($\times \sqrt{2}$)	4 –	14 14	28 30	50 50	78 86	124 124	172 180	242 242

D3S　Very awkward spud squares

(1) 24 yards

(2) 62 yards

(3) 510 yards

There are many different patterns giving the maximum journey, which has a length of $(n^3 - n)$ when n is odd, and of $(n^3 - 2)$ when n is even.

The diagrams in Fig. 43 show an easy way of writing out a maximum-length pattern for even n and (slightly differently) for odd n. For odd n, you can always get the maximum length without filling the central space, as the 3×3 pattern shown indicates: the journey is 24 yards without no. 9.

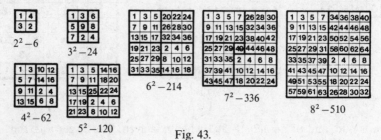

Fig. 43.

D4S　Rook's tours

Answers are shown in Fig. 44. None of the patterns is unique.

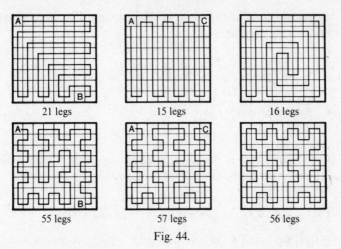

21 legs   15 legs   16 legs

55 legs   57 legs   56 legs

Fig. 44.

D5S   Shorter for Clancy

The route shown in Fig. 45 has a length of 38, and only 11 turns. It is not the only pattern.

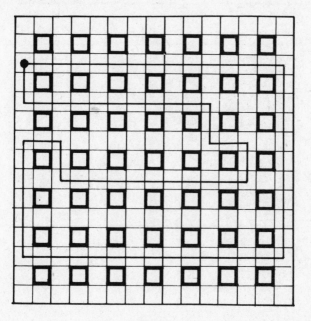

Fig. 45

My best results for the same problem with different numbers of houses in a square are:

Houses	9	16	25	36	49	64	81	100	121	144	169	196	225	256
length	8	16	22	34	38	50	64	80	86	100	122	142	150,	166
turns	3	7	7	11	11	15	15	19	19	23	23	27	27	31

D6S   Square cascades
(1) 42 ways,
(2) 24,024 ways.

If $n!$ ('factorial n') is the product of all the integers from 1 up to n, it appears that the number of ways of cascading a rectangle $a \times b$ is

$$(ab)! \times \frac{1!2!3!4!\ldots(a-1)!}{b!(b+1)!(b+2)!\ldots(b+a-1)!}$$

So the number of ways increases very rapidly. A 5×5 square has 701,149,020 ways to cascade it.

E Shapes and Shapes

E 1 Chequerboard cutting

(1) Divide a 6 × 6 chequerboard, by cutting along the lines only, into 6 pieces, each of which has
  (i) the same number of squares in it
 (ii) the same perimeter
(iii) the same amount of the original chequerboard-edge.
There are two solutions, with different piece-perimeters.

(2) Now divide the same board into 3 pieces, to satisfy the same three conditions, so that the perimeter of each piece is as *long* as possible.

(3) Again, into 3 pieces, same conditions, but making the piece-perimeter as *short* as possible.

(4) What is the smallest rectangular (not necessarily square) chequerboard which you can cut along the lines into two pieces, with the same number of squares in each, but with one perimeter *twice* as long as the other.

E2 Square dissecting

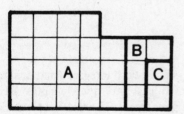

Fig. 46 .

45

Fig. 46 shows how to demonstrate $4^2 + 3^2 = 5^2$, in a 3-piece dissection. In how few pieces can you demonstrate:

    (1) $12^2 + 5^2 = 13^2$?
    (2) $15^2 + 8^2 = 17^2$?
    (3) $12^2 + 1^2 = 9^2 + 8^2$?
    (4) $13^2 + 4^2 = 11^2 + 8^2$?

(Two squares on one side of an equation must be joined neatly to make one straight edge.)

It is more elegant not to turn pieces over.

E3   Squared squares

The problem here is to fill an integral-sided square N^2 with as *few* smaller integral-sided squares as possible. If N is not a prime number, the sides of at least two squarelets must have lengths with no common factor higher than 1. Thus, in Fig. 47(B), 2 and 1 have no common factor above 1.

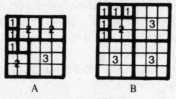

A   B

Fig. 47.

5^2 and 6^2 cannot be squared in fewer squarelets than shown in Fig. 47.

In how few squarelets can you square:

      13^2?
      17^2?
      23^2?
      40^2?
      41^2?

E4   Maximin squared squares

The problem is how to fill an n^2 square with integral-sided squarelets. If N is not prime, the sides of at least two squarelets must have lengths with no common factor higher than 1. The smallest squarelet is to be as large as possible, and it doesn't matter how many squarelets are used.

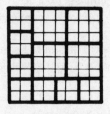

Fig. 48.

As Fig. 48 shows, 8^2 can be squared 'min 2'. No smaller N^2 can be.
What is the smallest N^2 which can be squared?

min 3
min 4
min 8
min 12
min 16

E5   Minor squared squares

Now we are trying to fill a large square N^2 with as few squarelets
as possible; at least two squarelets must have sides prime to each
other, and in addition, every squarelet's side must be *less* than $\frac{1}{2}N$.
Sometimes, but not always, this last rule means that we can't square
an N^2 in as few squarelets as we could without it. In how few squarelets can you, with this extra 'Minor' rule, square 11^2; 16^2; 40^2; 41^2?

E6   Triangled triangles

The problem of triangling triangles is not very different from that of
squaring squares. An equilateral triangle of side N has an area of N^2
units. Often a squared-square arrangement can be transformed into
a triangled-triangle arrangement; figure 49 shows two instances of
11^2 in 11 (one of them 'min 2'). And it is thus often pos. ible to

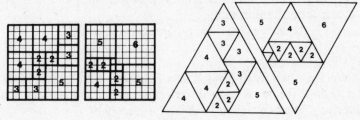

Fig. 49.

triangle N^2 in as few trianglets as it is to square N^2 in squarelets.

(1) What values of N^2 up to 41^2 cannot be triangled in as few as they can be squared in?

(2) Can you find instances of squared-square arrangements which *cannot* be transformed into a triangled-triangle arrangement, and vice versa?

E7 *Differently-squared squares*

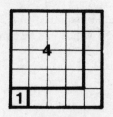

Fig. 50.

The problem here is to fill an integral-sided N square *as full as possible* with smaller integral-sided squarelets, each of a different size. For small N, you cannot do better than use a 1^2 and an $(N-1)^2$ squarelet, leaving $(2N-2)$ unfilled. The best you can achieve for $N = 5$, for instance, is 8 unfilled, as Fig. 50 shows.

How small an unfilled area can you leave after differently-squaring: 15^2; 16^2; 21^2; 41^2?

E8 *Progresso-perfect farm*

A progresso-perfect farm is a rectangle (with integral sides) subdivided into fields (smaller rectangles, also with integral sides), which is both:

(i) 'perfect'—i.e. no group of fields (apart from the whole farm) makes up a rectangle and

(ii) 'progressive'—i.e. it contains n fields with areas of 1, 2, 3 . . ., n square spoofs respectively.

What is the smallest possible progresso-perfect farm?

E9 *Slab and slablets*

Figure 51 shows a slab divided into 7 slablets of different sizes. In fact each of the 16 different dimensions, of slab and slablets, is a different whole number; so I call it a 'proper' slab.

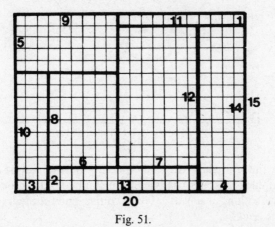

Fig. 51.

What is the smallest possible proper slab? The number of slablets into which it is divided may be 7 or more or fewer, as you choose. By 'smallest' I mean smallest in area.

E10 *Acungulation*

An acungle, let us say, is a triangle with each of its angles *smaller* than a right angle. Into how *few* acungles can you divide
   (1) a right-angled triangle?
   (2) an obtuse-angled triangle?
   (3) a square?
   (4) a 2 × 1 rectangle?
   (5) a regular pentagram (Fig. 52)?

Fig. 52.

E11   Obtuse/acute isosceles

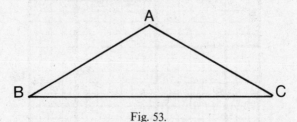

Fig. 53.

Divide the triangle in Fig. 53, which is obtuse and isosceles, into acute-angled isosceles triangles. You can make the angle at A what you like, so long as it is over 90° of course, and the angles at B and C must be equal.

The fewer acute-angled isosceles you end up with the better, and better still if any of them are equilateral.

E12   Circles in circles

Fig. 54.

The smallest circle that will contain 7 unit-radius circles has a radius of 3. We can say C(7)=3 (see Fig. 54).

What are C(11) and C(12)?

E13   Obtungulating polygrams

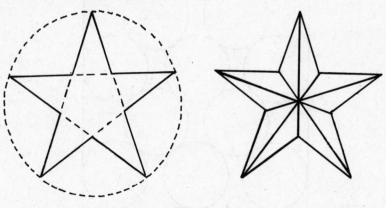

Fig. 55.

Left in Fig. 55 is a 5/2 polygram, or '5/2-gram' for short. If you space p points equally round a circle, and join each to the q'th point away, the shape of the outside lines is a p/q-gram. A $p/1$-gram is a regular p-gon. For q greater than 1 (but less, as it must be, than $p/2$), a p/q-gram has $2p$ sides.

The problem here is obtungulating polygrams—dividing them up into as few obtuse-angled triangles as possible.

(1) First consider $p/1$-grams; the equilateral triangle, the square, etc. In which cases can you *not* obtungulate into p obtungles? in which cases can you obtungulate in fewer?

(2) Next, p/q-grams with $q = 2$ or more. These are always obtungulable in $2p$, by joining each vertex to the centre, as in the second diagram in Fig. 55. Among polygrams with p ranging up to say 12, in how many cases can you improve on the figure of $2p$?

E14   Pennies in a square
Assuming the radius of a penny to be 10 gm, Fig. 56 (over) shows that 10 pennies can just be fitted into a square of 68 gm side.

Can you do better? Can you fit 10 pennies into a smaller square than that?

E15   Shapes in shapes
This is about fitting the largest possible regular figure of one shape

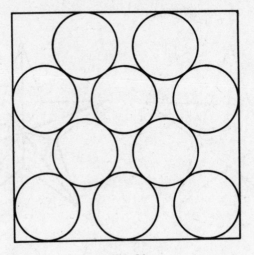

Fig. 56.

(triangle, square, hexagon, or circle) inside another, so that the ratio of the area of the inner figure to that of the outer is as large as possible.

'3 in 0', for instance, is $(3\sqrt{3})/(4\pi)$, because the largest equilateral triangle which can be fitted in a circle takes up $(3\sqrt{3})/(4\pi) = \cdot413$ of its area.

(1) Which is bigger, 3 in 4 or 4 in 3?

(2) 3 in 6 in 3, or 6 in 3 in 6?

(3) Using 3, 4, 6 and 0 once each, make A in B in C in D as large as possible.

(4) Similarly, make A in B in C in D in A as large as possible.

E16   Folding cubes and octahedra

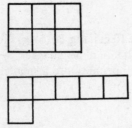

Fig. 57.

(1) Neither of the two 'hexomino' patterns in Fig. 57 can be folded along the lines so that the 6 squares form the 6 faces of a cube. But there are 33 other hexomino patterns (ignoring reflections); how many of them *can* be folded into a cube?

(2) Similarly, neither of the 'octiamond' patterns shown can be folded into an octahedron. How many of the other 64 octiamond patterns can?

E   Solutions

E1S   Chequerboard cutting
Fig. 58 shows the answers. Those for (2) (3) and (4) are not unique.

(1) peri 12   (1) peri 14   (2) peri 26   (3) peri 18

(4) 5 × 6: peris 16 and 32

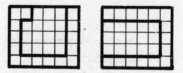

Fig. 58.

E2S   Square dissecting
Fig. 59 shows solutions for $15^2 + 8^2 = 17^2$ in 4 pieces, and for the others in 3 pieces.

The method shown for $12^2 + 5^2 = 13^2$ can be extended to solve in 3 pieces all cases of the type $A^2 + B^2 = (A+1)^2$, and that for $15^2 + 8^2 = 17^2$ to solve in 4 pieces all cases of the type $(2M-1)^2 + N^2 =$

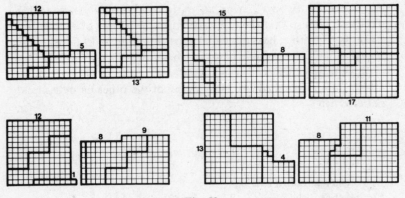

Fig. 59.

$(2M + 1)^2$. Leaving these aside, I can also solve in 4 pieces the cases where C (the single big square) is 29, 53, 73, 97, 109, 137, and in 5 pieces those where C is 65 (with 33, 56), 85 (with 36, 77), 89, and 125.

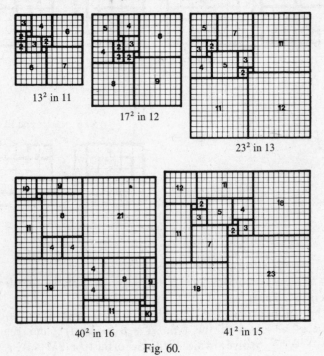

13^2 in 11

17^2 in 12

23^2 in 13

40^2 in 16

41^2 in 15

Fig. 60.

As for $A^2 + B^2 = C^2 + D^2$, the following cases are soluble in 3 pieces:

8,1 = 7,4
9,2 = 7,6
11,2 = 10,5 (and similarly 23,2 = 22,7; 39,2 = 38,9, etc.)
12,1 = 9,8 (and similarly 30,7 = 25,18; 56,17 = 49,32 etc.)
13,4 = 11,8

And I have solved all other cases where the total area is less than 700 in 4 pieces. I should be delighted to hear of improvements.

E3S *Squared squares*

Fig. 60 shows the answers.
The following table summarizes the best results I have found so far:

Number of Squarelets	N
6	3
7	4
8	5
9	6 and 7
10	8 and 9
11	10–13
12	14–17
13	18–23
14	24–29
15	30–39 and 41
16	40 and 42–50
17	51–65 and 68 and 70
18	66, 67, 69 and 71–82
19	83–107
20	108–127 and 129–131

If N(S) is the lowest *odd* value of N which can't be squared in fewer than S squarelets, the formulae
$$N (2k+1) = 4F (k-2) - 1 \text{ and } N (2k) = 2F (k-1) - 1,$$
wherein $F(n) =$ the *n*th Fibonacci number, produce the right results so far as I can see up to S = 20, except that N = 3 doesn't require 7 squarelets and N = 41 doesn't require 16.

E4S *Maximin squared squares*

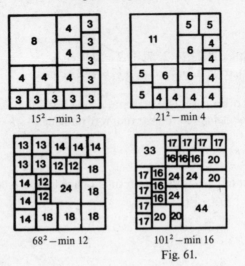

Fig. 61.

Fig. 61 shows the answers. Notice how 42^2 and 101^2 are related to 21^2.

The following table shows the lowest N^2 I have found for all minimum squarelet sizes from 2 up to 16:

Min	2	3	4	5	6	7	8	9	10	11	12	13	14	15	16
N	8	15	21	29	32	43	42	54	64	63	68	92	86	96	101

E5S *Minor squared squares*

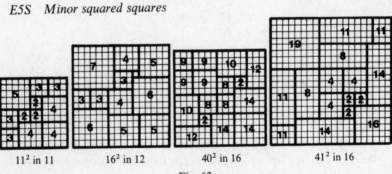

Fig. 62.

Fig. 62 shows the answers. Of these, 41^2 (for which there are several other patterns) is the only one where the 'Minor' rule seems to force an extra squarelet. The other cases (for N from 8 to 41) where, so far as I have got, extra squarelets are needed for a 'Minor' solution, are 8, 9, 12 (2 more) 13, 15, 17, 21, 22, 23, 26–29, 33, 34, 36, 38, 39, 41; but I expect this list could be shortened with more work.

E6S   Triangled triangles
(1) N = 13, 17, 22, 23, 29, 38, 39, 41.
(2) Some examples are shown in Fig. 63.

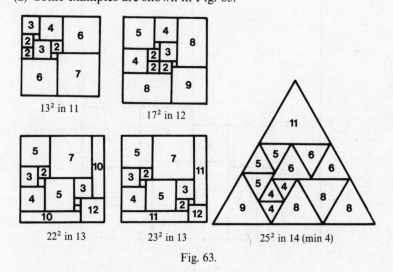

13² in 11   17² in 12   22² in 13   23² in 13   25² in 14 (min 4)

Fig. 63.

Notes   (1) The best 'Maximin' results obtainable for N up to 41 seem to be the same for triangles as for squares.

(2) I have so far found no instance where a result for an N^2 triangle is *better* than the result for an N^2 square.

E7S   Differently-squared squares
Fig. 64 shows answers.
15^2 seems to be the smallest case where you can improve on 1^2 and $(N-1)^2$, and 16^2 the largest case where you cannot. The following table shows my best results so far:

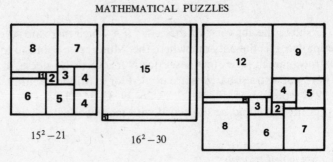

$15^2 - 21$ $16^2 - 30$

$21^2 - 12$

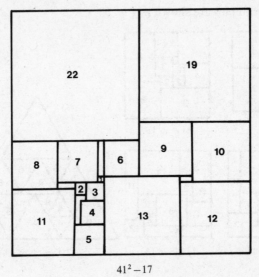

$41^2 - 17$

Fig. 64.

N	15	16	17	18	19	20	21	22	23	24	25	26	27	28	29	30
leaving	21	30	29	20	25	30	12	19	24	17	17	13	18	14	21	14

N	31	32	33	34	35	36	37	38	39	40	41	42	43	44	45
leaving	19	15	15	20	20	21	16	22	16	16	17	26	22	17	18

Dudeney's puzzle 'Lady Isabel's Casket' is based on N80, with 40 left unfilled.

E8S   Progresso-perfect farm

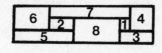

Fig. 65.

36 square spoofs, which can be laid out as shown in Fig. 65.

The area must be the sum of the numbers from 1 to n, i.e. $(n(n+1))/2$. A perfect farm must have more than 4 fields. Can $n=5$? The area is 15, which must be laid out as 5×3. Where is the 5-field to go? It must make the farm imperfect.

$n=6$ will not work; perfection is impossible with 6 fields.

$n=7$ gives an area of 28 i.e. 14×2 or 7×4. A width of 2 clearly cannot yield perfection. And where is the 7-field to go in a 7×4 shape?

So $n=8$, with area 36, is the least possible; and that is practicable, as shown.

E9S   Slab and slablets

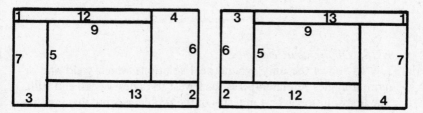

Fig. 66.

A slab 16×8 can be 'properly' divided as shown in Fig. 66. A proper slab can be made up of 7 or more slablets, or of 5, but 6 is not possible, nor is any number less than 5.

E10S   Acungulation

(1) and (2) 7 (Figs. 67 A and B).

(3) 8 (Fig. 67C, where the dotted lines show limits of position for A and B)

(4) 8 (Fig. 67D, showing all can be isosceles and two equilateral)

(5) 6 (Fig. 67E, showing possible angles)

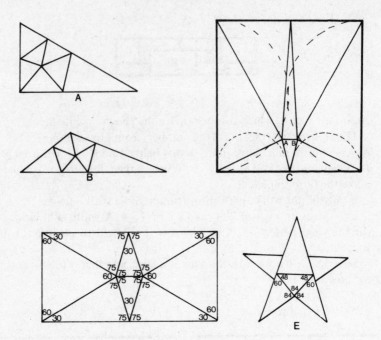

Fig. 67.

E11S   Obtuse/acute isosceles

Fig. 68 shows the only symmetrical solutions which yield one or more triangles equilateral. 7 is the fewest triangles we can end with. The first solution is best in terms of the problem as put, having two equilaterals.

E12S   Circles in circles

C(11) is 3·924: C(12) is 4·030.

In Fig. 69, A shows the arrangement for 11 circles. n circles will just go round the inside of a larger circle of radius $(1 + \operatorname{cosec}(180°/n))$, so 9 will fit in $(1 + \operatorname{cosec} 20°) = 3·924$, and 2 others will just fit in the middle as shown.

B shows the arrangement for 12 circles. It has 3 axes of symmetry.

E13S   Obtungulating polygrams

(1) Only the square needs more (see A in Fig. 70). The general

method shown in B–H yields exactly p obtungles for all other p-gons. I don't think any p-gon can be obtungulated in fewer than p.

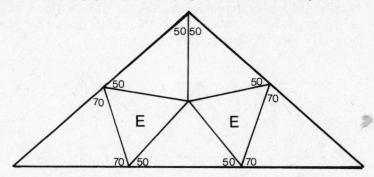

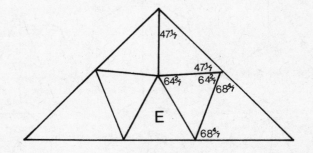

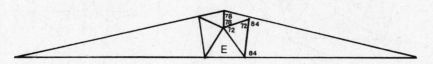

Fig. 68.

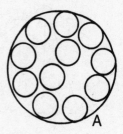

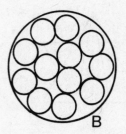

Fig. 69.

(2) I–L in Fig. 70 show 5/2 in 8, 6/2 in 10, 7/3 in 8, and 10/2 in 10. Also possible are 9/2 in 8, 11/2 in 12, and 12/2 in 12. I should be surprised if further improvements are not possible. (Incidentally, comparing L with C shows how to obtungulate in *n* all *n*/2-grams where *n* is even and greater than 8.)

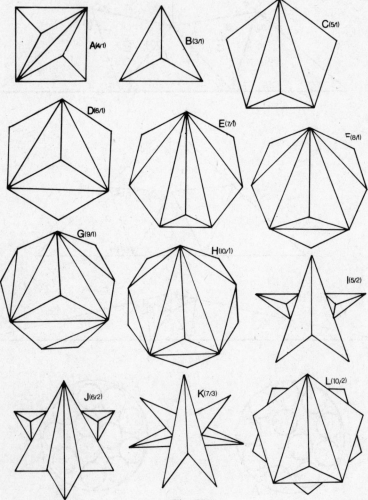

Fig. 70.

E14S   Pennies in a square

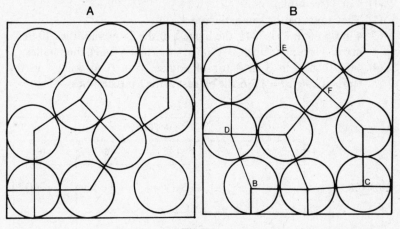

Fig. 71.

Fig. 71A shows how to fit them into a square with sides 10 $(4 + \sqrt{2} + 4\sqrt{2})$ or about 67·67 gm. The co-ordinates of A are $[20 + 5(\sqrt{2} + \sqrt{2} + 4\sqrt{2})]$, $[20 + 5(\sqrt{2} + 4\sqrt{2} - \sqrt{2})]$.

Fig. 71B fits them into an even smaller square, with sides about 67·45 gm. Co-ordinates are: B; 17·47, 10   C; 57·45, 10·85   D; 10, 28·56   E; 27·59, 57·45   F; 42·52, 44·15.

How big a square is needed to take *n* pennies is often a teasing question. My best results are:

n	square side	n	square side	n	square side
1	20	11	70·32	21	
2	34·14	12	71·45	22	94·64
3	39·32	13	74·64	23	97·32
4	40	14	77·32	24	98·63
5	48·28	15	78·63	25	100
6	53·29	16	80	26	
7	57·32	17		27	104·80
8	58·63	18	86·57	28	
9	60	19		29	
10	67·45	20	89·78	30	109·09

E15S   Shapes in shapes
(1) 4 in 3 (see table).
(2) They must be the same. Each is $\frac{1}{3}$.
(3) 4 in 0 in 6 in 3 is ·3846, the biggest of the 24 possibilities. (0 in 4 in 6 in 3 is runner-up at ·3766. 6 in 4 in 3 in 0 is ·1434, the smallest).
(4) 4 in 0 in 6 in 3 in 4 (or of course, 0 in 6 in 3 in 4 in 0, etc) is $\frac{2}{3}(2-\sqrt{3})=\cdot 1786$, the biggest of the 6 possibilities.

		OUTER			
		3	4	6	0
INNER	3	1	$2\sqrt{3}-3=\cdot 464$	$\frac{1}{2}=\cdot 500$	$\frac{3\sqrt{3}}{4\pi}=\cdot 413$
	4	$4(7\sqrt{3}-12)=\cdot 499$	1	$\frac{2\sqrt{3}+3}{9}=\cdot 718$	$\frac{2}{\pi}=\cdot 636$
	6	$\frac{2}{3}=\cdot 667$	$\frac{6\sqrt{3}-9}{2}=\cdot 696$	1	$\frac{3\sqrt{3}}{2\pi}=\cdot 827$
	0	$\frac{\pi\sqrt{3}}{9}=\cdot 605$	$\frac{\pi}{4}=\cdot 786$	$\frac{\pi\sqrt{3}}{6}=\cdot 907$	1

E16S   Folding cubes and octahedra
11 in each case. And there is a fuzzy correspondence between the two sets, comparing the patterns with the same letter as shown in Fig. 72.

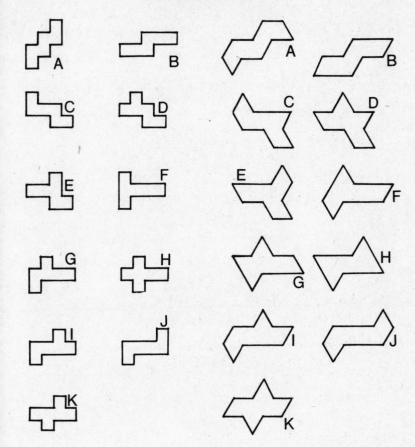

Fig. 72.

F   Games

F1   The best die

'Here are four dice,' said Eli; 'Each, as you see, is the usual cube shape. None is loaded. All roll true. We'll choose a die each, and roll for pennies. Higher number wins a penny each time'.

'What about ties?' I said.

'I've numbered each die differently, in such a way that ties are impossible. Come on. I'll let you have first choice of die'.

'Well', I said, 'I want the one whose average throw is the highest, and that'll be the one whose numbers add up to the largest total. I'll have that one'.

Eli chuckled: 'All the number totals are the same. Have a look'.

I did, and they were. Baffled, I chose one at random. Eli chose one. As I might have expected, Eli turned out to be winning twice as often, on average, as I.

How did Eli ensure this result, with a single-digit number (no zeroes) on each face of each die?

F2   The better Bowler

Fred and George between them took all 10 wickets in each innings of our 2-innings match against Downsea (who scored 60 runs in all, with no extras). So they were the only two contenders for the Pearson Pint, awarded for the best Uplands bowling performance. Fred's average (i.e. runs per wicket) in the first innings was lower than George's, and so it was in the second innings too. But before giving Fred the Pint, I totted up the combined figures and found to my surprise that George had the lower average for the whole match. So I regarded their performances as equal—especially as each had taken 10 wickets in all—and I drank the Pint myself.

All the averages were exact whole numbers. What were they?

F3   The best golfer

Maurice and Joris played the first 9 holes at Flixby, which consist of 3 par-3 holes, 3 par-4 holes, and 3 par-5 holes, in the annual competition for the Slaithwaite Sandwich. Each had 3 3's, 3 4's, and 3 5's in his score: and each had 3 birdies (1 under par), 3 pars and 3 bogeys (1 over par). At no hole did both have the same score. Maurice got a birdie at the par-4 2nd hole, but not at the par-4 6th or 7th.

Since they had the same total, I decided to award the Sandwich to whichever would have won if they had been playing a match, i.e. to the one who took the fewer strokes at more holes.

Who won the Sandwich?

F4   Soo, Hadoto and us

Peter Soo and Charlie Hadoto often play, for hours on end, a game in which each holds up simultaneously a hand with 1, 2, 3, 4 or 5 fingers extended. If the total of fingers extended by both is Prime, Peter wins a penny. If the total is Composite, Charlie wins a penny.

(1) Who is more likely to win in the long run, Peter or Charlie?

(2) Suppose you and I play the same game, but for different stakes, the amount handed over each time being not a penny but the same number of pennies as the total of fingers extended. Which side will you take—the chap who wins if the total is Prime, or the one who wins if it is Composite?

(You should of course assume that Peter, Charlie, you and I are all intelligent sensible people).

F5   Revolutionary championship

Each year on the umpteenth of Cuspidor, the same plonkteen teams meet for the Revolutionary Championship. Every day till the first of Corridor, each team revolutes competitively against another, till each team has played every other twice. Two points for a win, one for a draw. In the final Championship Table, ties are split by Prole Average—rather like the Football League, in fact.

Last year, Awdry wanted very much to win the Championship, and things went as well as they possibly could for them from the start. In fact, the joshteenth match put them into an unassailable position—bound to win the Championship. This they did, with Bawdry coming second.

This year, Bawdry didn't care about winning the Championship itself, but they did very much want to finish ahead of Awdry. Things went the best possible way for them from the start, and the quimpteenth match put them in a position when they were bound to beat Awdry.

If joshteen *plus* quimpteen *equals* 50, what number in English does plonkteen represent?

F6   The bridge odds
(1) Which are there more of, different deals or different auctions, at contract bridge? (Ignore misdeals and illegal calls).
(2) What is the shortest possible bridge auction?

F   Solutions

F1S   The best die
The four dice were (A) 993333
                   (B) 888222
                   (C) 777711
                   (D) 665544 (or 655554 or 666444 or 555555)

A beats B beats C beats D beats A, just 2 times out of 3 in the long run.

Having the number-total of each die the same ensures that the dice are equally 'good' in a game in which payment is proportionate to the *amount* by which one player beats the other each time. But with payment at a flat rate, dice cannot be ranked in a fixed order of goodness, and there is no 'best die' in such a set as this.

With a set of 3 dice, you can get a similar though not quite so striking effect. With
(A) 444411
(B) 333333
(C) 552222

A beats B $\frac{2}{3}$ of the time; B beats C $\frac{2}{3}$; C beats A only $\frac{5}{9}$.

F2S   The better bowler
Fred: 2 for 0 and 8 for 40; total 10 for 40.
George: 8 for 8 and 2 for 12; total 10 for 20.

We need to find integers p, q, x and y (each positive or zero) and a (an integer between 0 and 10) to give:

	1st Innings	2nd Innings	Overall
Fred $\left(\dfrac{\text{runs}}{\text{wickets}}\right)$	$\dfrac{ax}{a}$	$\dfrac{(10-a)y}{(10-a)}$	$\dfrac{ax+(10-a)y}{10}$
George $\left(\dfrac{\text{runs}}{\text{wickets}}\right)$	$\dfrac{(10-a)p}{(10-a)}$	$\dfrac{aq}{a}$	$\dfrac{aq+(10-a)p}{10}$

We know that (a) $x<p$; $y<q$;

               (b) $10y+a(x-y)>10p+a(q-p)$;

               (c) 10 divides both $a(x\text{-}y)$ and $a(q-p)$;

               (d) $a(x+q)+(10-a)(p+y)=60$ or some lower multiple of 10.

From (b), $a/(10-a)<(y-p)/(q-x)$. Since $a/(10-a)$ is positive, $y>p$ and $q>x$, or $y<p$ and $q<x$. Assume the former, since the effect is only on the order of the 2 innings. So, taking account of (a), $q>y>p>x$. So $a/(10-a)<1$. So $a<5$, and we can satisfy (c) only by:

    (1) $a=1$ or 3: $(y-x)$ and $(q-p)$ are multiples of 10; or

    (2) $a=2$ or 4: $(y-x)$ and $(q-p)$ are multiples of 5.

A little further work shows that with (1) the run-total must exceed 60, and that (2), with $a=2$, $p=1$, $q=6$, $x=0$, $y=5$ provides the only answer.

F3S   The best golfer

Maurice.

For each player, we need to fill Table 1. To satisfy the condition about 3 birdies, 3 pars and 3 bogeys, $b+3-c-d=c+3-b-d=a+d+a+b+c+d-3=3$. So $b=c$, $d=0$, $a+b=3$.

Par	Score		
	3	4	5
3	a	b	$3-a-b$
4	c	d	$3-c-d$
5	$3-a-c$	$3-b-d$	$a+b+c+d-3$

Table 1

So the table can be filled in in 4 different ways only (Table 2). And since the players' scores were different at all holes, and M had just one birdie at a par-4 hole, their scores were as in Table 3.

| | Score | | |
Par	3	4	5
3	3/2/1/0	0/1/2/3	0
4	0/1/2/3	0	3/2/1/0
5	0	3/2/1/0	0/1/2/3

Table 2

| | Score | | |
Par	3	4	5
3	M2 J1	M1 J2	0
4	M1 J2	0	M2 J1
5	0	M2 J1	M1 J2

Table 3

So M won 2 par-3 holes, 1 par-4, and 2 par-5; 5 holes to M, and 4 to J. So Maurice won.

F4S   Soo, Hadoto and us
(1) Neither. The chances favour each player in the proportion of 'Six Of One' to 'Half A Dozen Of The Other'
(2) You would be wise to be Mr Prime. You should win, on average, 1p every 28 hands; though Mr C will on average win 197 hands to Mr P's 195.

	C				
	1	2	3	4	5
1	P	P	C	P	C
2	P	C	P	C	P
3	C	P	C	P	C
4	P	C	P	C	C
5	C	P	C	C	C

P

Table 1

		C	
		$4(x)$	$5(1-x)$
P	$1(y)$	$P(xy)$	$C(1-x)y$
	$2(1-y)$	$Cx(1-y)$	$P(1-x)(1-y)$

Table 2

		C	
		$4(x)$	$5(1-x)$
P	$2(y)$	$C6xy$	$P7(1-x)y$
	$3(1-y)$	$P7x(1-y)$	$C8(1-x)(1-y)$

Table 3

For the first game, *Table 1* sets out the possible results. It shows that, for P, showing 1 'dominates' showing 3; that is to say, whatever C does, P cannot lose and may gain from showing 1 rather than 3. Similarly, for P, 3 dominates 5, and 2 dominates 4. Now consider C. For him, 5 dominates 3 which dominates 1, and 4 dominates 2. So P should always show 1 or 2, and C should always show 4 or 5. How often should each show his two numbers respectively? If C shows 4 a proportion x of the time, and P shows 1 y of the time, *Table 2* enables us to calculate P's average net gain per hand as

$$xy - (1-x)y - x(1-y) + (1-x)(1-y) = (2x-1)(2y-1).$$

Now if C makes x other than $\frac{1}{2}$, P can choose y more or less than $\frac{1}{2}$, so as to make his gain positive. But if C makes $x = \frac{1}{2}$, P's gain will be 0. Similarly P cannot afford to make y other than $\frac{1}{2}$, or C can adjust x to make P's gain negative.

So the wise strategy for each player is to spin a coin before each hand and show one of his numbers if it is heads and the other if it is tails. Then the chances are even.

Can it ever be wise to deviate from that strategy? If P does so, the chances remain even unless and until C also deviates. When P detects this, he may be able to deviate the other way and benefit— till C in turn detects this and reverts to basic strategy. Should this ever happen? I think not. There is perhaps at first sight a greater incentive to C to start a deviation pattern, for if he can get P to do so too the odds may move in the direction of the 14 to 11 in favour of C which apply if each player showed 1, 2, 3, 4 or 5 completely at random. Or either player may fancy his chances in the psychological poker-like situation which results if the choice is made directly each time, and not by some coin-tossing process. But, assuming each player is intelligent enough to make the same calculations that the other makes, it appears that, if there is *any* line of reasoning that might encourage either player to deviate, the other would be aware of that same line of reasoning and would protect himself by sticking to the basic strategy.

My own conclusion is that two sensible and intelligent players should each adhere to their basic strategy.

As to the second game, for P there are no 'dominances' ab initio. But C has the same dominances as before—so will never show 1, 2 or 3—and *now*, for P, 3 dominates 1, and 1, 2 and 3 all dominate 4

and 5. So the table reduces to *Table 3*. We can now calculate that
P's net gain is

$$y(-6x+7-7x)+(1-y)(7x-8+8x)=\tfrac{1}{28}-28(x-\tfrac{15}{28})(y-\tfrac{15}{28}).$$

Reasoning as before, we conclude that:
  C should show 4 $\tfrac{15}{28}$ of the time, and 5 $\tfrac{13}{28}$;
  P should show 2 $\tfrac{15}{28}$ of the time, and 3 $\tfrac{13}{28}$;
  P's expected net gain is $\tfrac{1}{28}$ p per hand.
It is interesting to note that C will win *more often*, however. In every
$784(=28^2)$ hands, he expects to win $394(=15^2+13^2)$, while P expects
to win $390(=13.15+15.13)$.

F5S The revolutionary championship
22. After each of t teams has played x matches, the strongest position
for A is to have scored $2x$ points, with the other teams sharing the
remaining $tx-2x$ points as equally as possible, i.e. averaging
$x(t-2)/(t-1)$ apiece. Unless that is an integer, some will have
scored $[x(t-2)/(t-1)]$ and some $[x(t-2)/(t-1)]+1$. If, at that
point, A's position is to be proof against scoring no further points
itself and one of the teams with $[x(t-2)/(t-1)]+1$ scoring 2
points in each of its $j(t-1)-x$ remaining games (j being the number
of times each team plays each other),

$$2x+0>\left[\frac{x(t-2)}{t-1}\right]+1+2j(t-1)-2x.$$

The 'integer equal to or just less than' feature represented by $[\ ]$
makes this a bit difficult to work out as a formula for x in terms of
t and j. But an empirical approach shows that, for $j=2$, x is the
integer *nearest* to $(4t-3)/3$.

As for this year, B's position is strongest if, after y matches, B has
scored $2y$ and A has scored 0. To be proof against A scoring 2 points
in each of the remaining $j(t-1)-y$ games, while B scores 0 in each
of them, we need

$$2y+0>0+2j(t-1)-2y$$
i.e. $2y>j(t-1)$, or, with $j=2$ (as here),
  $y>t-1$, i.e. $y=t$.

Since $\dfrac{4t-3}{3}+t=50$, approximately, $7t-3=150$: $t=\dfrac{153}{7}$: $t=22$.

So plonkteen is 22; and joshteen is 28, quimpteen being another word for 22.

F6S   *The bridge odds*

(1) The number of different auctions is greater. It is about $2\frac{1}{2}$ million million million times the number of different deals.

(2) 7NT, Double, Redouble: 3 bids only.

The number of deals is, fairly clearly, $52!/(13!)^4$. We can use Stirling's formula, that $n!$ is approximately $\sqrt{2\pi n} \times (n/e)^n$, to evaluate this as about $5{\cdot}35 \times 10^{28}$.

The number of auctions needs a little analysis. An auction consists of a number of bids (B) of a certain number of a suit (or No Trumps), possibly interspersed with other calls, of 3 types: a pass (P), a double (D), and a redouble (R). The rules for all these are: a B must be superior to the previous B (in the ascending order

$$1C, \ 1D, \ 1H, \ 1S, \ 1NT, \ 2C, \ 2D \ldots 7NT):$$

a P can be made at any time, and the auction stops after 3 successive P's (except that, at the very beginning, there may be PPPP or PPPB): a D can only follow a B, either immediately or after 2 P's: a R can only follow a D, either immediately or after 2 P's.

Any two successive B's in an auction are therefore linked in one of the following ways:

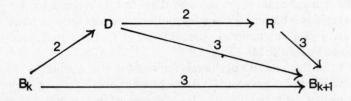

The numbers on the lines show the P alternatives (0 or 2 passes before a D or R; otherwise 0, 1 or 2 passes). So there are 21 ways of getting from any B to the next.

If we name 1 Club B_0, 1 Diamond B_1 etc., up to 7 No Trumps as B_{34}, calculation shows that the number of different auctions ending with B_n is 4.22^n, and other formulae as follows:

Call	No. of auctions ending there	No. of auctions ending there or sooner
B_n	4.22^n	$\frac{1}{3}(16.22^n - 1)$
$B_n + D$	8.22^n	$\frac{1}{3}(40.22^n - 1)$
$B_n + D + R$	16.22^n	$\frac{1}{3}(4.22^{n+1} - 1)$

Since 7 No Trumps, the highest bid, is B_{34}, the number of possible auctions is $(4.22^{35} - 1)/3$, or about $1{\cdot}285 \times 10^{47}$.

And $\dfrac{1{\cdot}285 \times 10^{47}}{5{\cdot}354 \times 10^{28}} = 2{\cdot}4 \times 10^{18}$ approximately.

G   Cubes and Bricks

G1   Edge-magic cube

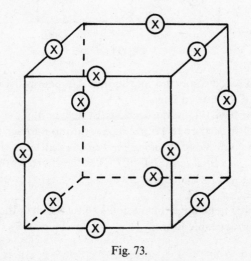

Fig. 73.

Fig. 73 shows a cube, with a label (*x*) on each edge. Label each edge with a different whole number (a positive one: no zeros), so that the numbers on the 3 edges round each vertex add up to the same 'V-sum', and the numbers on the 4 edges round each face add up to the same 'F-sum'.

If you cannot do it with the numbers 1–12, do it with the smallest set of different numbers possible.

G2   Flies over cube
Fig. 74 shows a solid cube, floating freely in space. Each edge is marked with a solid line, and there is a number at the midpoint of

77

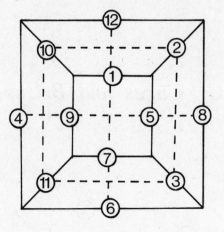

Fig. 74.

each edge. Each face (including the invisible one) is divided into 4 equal squares by dotted lines.

My 3 tame flies, Ed, Fred and Jed each start at number 1 and visit each number in turn up to 12 and then return to number 1. Ed takes the shortest route possible *along the edges*. Fred takes the shortest route *along the dotted lines*. Jed just takes the shortest route.

If each edge is 1 metre long, you will find Ed has to travel 24 metres, Fred 18, and Jed 15·07.

Is it possible to arrange the numbers so that each of the 3 flies has the longest practicable journey?

G3   *Bricks and cubelets*

Anne, Fran, and Jan each have a brick, painted red all over. Each saws her brick into one-inch cubelets, some of which of course are unpainted and some are painted (on one or more faces). Each finds that the number of her painted cubelets is exactly the same as the number of her unpainted.

'I have twice as many cubelets as Jan,' says Fran. 'And I,' says Anne, 'have as many as you both have together'.

What were the dimensions of each of their bricks?

(A 'brick', to these young ladies, is a rectangular parallelepiped, however unlike an ordinary building brick its proportions).

G4   Cubes and cubelets

I have 3 wooden cubes, whose outsides are painted blue all over. I saw them up neatly into one-inch cubelets, which of course are of two kinds; painted (on one face or more), and unpainted. The total number of painted cubelets turns out to be very nearly the same as the total of unpainted. In fact there is a difference of just 1 between the totals.

How big were my 3 original cubes?

G5   A cubeful of cubelets

(1) You cannot divide a square up exactly into 2 squarelets, but you can into 4, of course. For what values of n is it possible to divide a square into precisely n squarelets?

(2) And for what values of n is it possible to divide a cube into n cubelets? (Squarelets and cubelets may be of the same or different sizes, as convenient).

(3) What is the smallest number of cubelets with integral sides into which you can divide an n^3 cube, taking n as 3, 5, 7, 11 in turn?

G6   Triangles in cubes

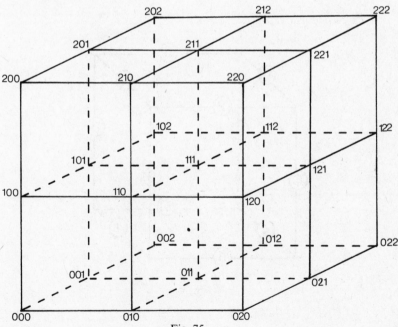

Fig. 75.

Fig. 75 shows 27 points, numbered for easy reference, in a regular cubical array.

(1) Can you find any equilateral triangles with each vertex at one of the points? If so, how many?

(2) Can you find any regular tetrahedra with each vertex at one of the points? If so, how many?

G7   Red and blue cube

Imagine a cube, divided into 27 equal transparent hollow mini-cubes. Can you put a blue ball or a red ball inside each mini-cube in such a way that you don't have 3 balls of the same colour in a straight line in any direction?

The answer is no. But you can put a red or blue ball in every minicube except the one in the very centre (the one which doesn't touch the surface of the cube), without having 3 of the same colour in line. How?

G   Solutions

G1S   Edge-magic cube

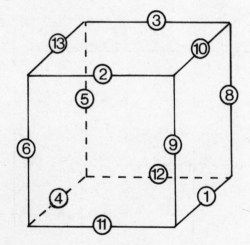

Fig. 76.

If N is the total of the 12 numbers we use, V is the V-sum, and F the F-sum, it is not hard to see that if we add up all the numbers round each of the 8 vertices in turn the total will be $8V = 2N$, for each edge has been counted twice. Similarly $6F = 2N$. So $N = 4V = 3F$, and N must be a multiple of 12. Now the numbers 1 to 12 add up to 78. So 84, the next higher multiple of 12, is the lowest practicable N.

So the numbers 1–13, omitting 7, are the set to try first. And it is possible with them, as shown in Fig. 76.

That arrangement, reflexions and rotations apart, is unique.

G2S   Flies over cube
Yes. Fig. 77 shows one way to do it.

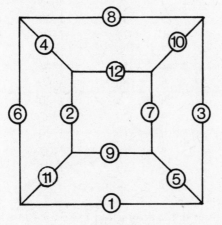

Fig. 77.

This gives Ed 30 metres, Fred 24 and Jed 20·484.

G3S   Bricks and cubelets
Jan's was $8 \times 10 \times 12$ in. $= 960$ cu. in.
Fran's was $6 \times 10 \times 32$ in. $= 1920$ cu. in.
Anne's was $5 \times 18 \times 32$ in. $= 2880$ cu. in.

If a brick is $(x+2)$ in. $\times (y+2)$ in. $\times (z+2)$ in., it yields a total of $(x+2)(y+2)(z+2)$ cubelets, of which xyz are unpainted. So

$$\tfrac{1}{2}(x+2)(y+2)(z+2) = xyz, \text{ or } x = \frac{2(y+2)(z+2)}{(z-2)y-2(z+2)}$$

We can safely assume $x \geqslant y \geqslant z$, and quite easily by trial find the 20 possible solutions:

$z+2$	$y+2$	$x+2$	$(z+2)(y+2)(x+2)$
5	13	132	8580
5	14	72	5840
5	15	52	3900
5	16	42	3360
5	17	36	3060
5	18	32	2880*
5	20	27	2700
5	22	24	2640
6	9	56	3024
6	10	32	1920*
6	11	24	1584
6	12	20	1440
6	14	16	1344
7	7	100	4900
7	8	30	1680
7	9	20	1260
7	10	16	1120
8	8	18	1152
8	9	14	1008
8	10	12	960*

And the only three in the proportion 1:2:3 are those marked*.

G4S   Cubes and cubelets
11 in., 8 in., and 4 in.

If you cut up an n^3 cube, you obviously get $(n-2)^3$ unpainted cubelets and $[n^3 - (n-2)^3]$ painted. So the 'excess' of painted over unpainted is $[n^3 - 2(n-2)^3]$, which may of course be positive or negative. As an alternative to a horrid cubic equation, note that only a small cube yields a positive 'excess'. If $n = 11$ or more, the excess is negative. For $n = 11$, it is -127, and for $n = 12$ it is -272. The biggest positive excess is 93, when $n = 7$. So the cubes must be 11 in. and two smaller ones with positive excesses totalling 126 or 128. Trial shows they can only be 8 in. (excess 80) and 4 in. (excess 48).

G5S   *A cubeful of cubelets*

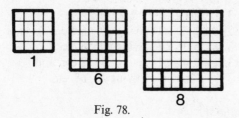

Fig. 78.

(1) All values except 2, 3, and 5.

(2) For $n = 1, 8, 15, 20, 22, 27, 29, 34, 36, 38, 39, 41, 43, 45, 46, 48, 49, 50, 51, 52, 53$; and for all values from 55 onwards.

(3) 3^3 into 20: 19 1s and a 2.

5^3 into 50: 42 1s, 7 2s and a 3.

7^3 into 71: 51 1s, 15 2s, 4 3s and a 4 (see Fig. 79 top).

11^3 into 125: 90 1s, 15 2s, 15 3s, 4 5s and a 6 (see Fig. 79 bottom).

(1) You can clearly replace any unit squarelet by a square foursome of squarelets, increasing the squarelet total by 3. By repeating this, you can develop the 3 patterns of figure 78 into

$$1, 4, 7, 10, 13 \ldots$$
$$6, 9, 12, 15 \ldots$$
$$8, 11, 14 \ldots$$

and thus achieve any number of squarelets except 2, 3 and 5.

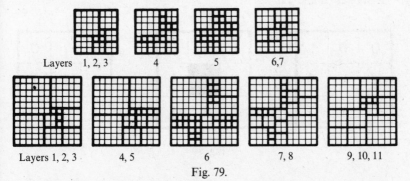

Fig. 79.

(2) Similarly, you can replace any unit cubelet by a cubic eightsome, increasing the cubelet total by 7. The seven basic solutions we need are:

1^3 in 1; giving 1, 8, 15. 22, etc.

3^3 in 20 (1×2, 19×1) giving 20, 27, 34 etc.

4^3 in 38 (1×3, 37×1) giving 38, 45 etc.

6^3 in 39 (6×3, 3×2, 30×1) giving 39, 46 etc.

6^3 in 49 (4×3, 9×2, 36×1) giving 49, 56 etc.

6^3 in 51 (5×3, 5×2, 41×1) giving 51, 58 etc. and

6^3 in 61 (3×3, 11×2, 47×1) giving 61, 68, etc.

G6S Triangles in cubes

(1) There are 80

(2) There are 18

In each of the 8 small cubes (such as 000, 010, 011, 001, 100, 110, 111, 101) there are 2 tetrahedra (such as 000, 011, 101, 110), each with 4 equilateral triangles as faces. And the full cube provides two double-sized tetrahedra (such as 000, 022, 202, 220), giving 8 more triangles. Total 18 tetrahedra and 72 triangles.

In addition, there are 8 triangles (such as 001, 120, 212) with centres at point 111, which are not faces of tetrahedra with all 4 vertices on a regular cubical array (even a larger one).

The sides of the 3 lots of triangle are $\sqrt{2}$, $2\sqrt{2}$, and $\sqrt{6}$ respectively.

G7S Red and blue cube

All the cube's faces show the same pattern: see figure 80. This is the only solution (apart from exchanging the reds and blues—represented here as 0s and 1s).

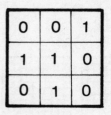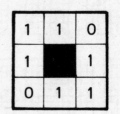

Fig. 80.

H How Many?

H1 How few flyovers?

If you aren't concerned about distance, you can connect 4 towns, each to each, by motorways without any flyovers. But, to interconnect 5 towns, you must have at least one flyover. In Fig. 81, you can't join CE without crossing BD or AB or AD.

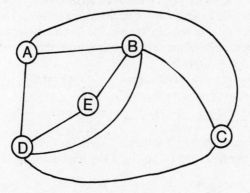

Fig. 81.

How many flyovers must you have to link 13 towns? (All flyovers are to be separate: no multi-level junctions)

H2 The committee function

'I very much welcome', said the Chairman, 'the increase in the Committee's size. This will enable us to face the problems of the future with substantially greater . . . '

'What it *will* do' said the Secretary to himself, 'is increase the number of ways the Committee can agree or disagree by—let's see—exactly 17,007'.

What was the Committee's size, before and after the increase? (A committee of 2 can 'agree or disagree' in 2 ways: AB or A/B; one of 3 in 5 ways: ABC or A/BC or B/AC or C/AB or A/B/C. And so on).

H3   Name-squares

Fig. 82 shows Ada's first shot at putting letters in a 3 × 3 grid so as to

A	D	A
A	D	A
D	A	D

Fig. 82.

let her spell her name in as many ways as possible. (She starts at any A: then moves, up or down or left or right (not diagonally), to an adjacent D: then to an adjacent A. She can use the same letter twice if she wants.) As you see, she scored 21: 2 starting with each of the top As, 5 from each of the side As, and 7 from the bottom A.

(1) Can she do better?
(2) How many can Anna score (in a 4 × 4 grid)?
(3) And Anona (in a 5 × 5 grid)?

H4   Name-hexes

In Bob's name-hex (Fig. 83), he can spell out his name, proceeding letter by letter from cell to touching cell *without* using the same cell

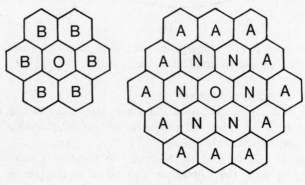

Fig. 83.

twice, in just 30 different ways. For he can start at any of the 6 Bs: must then move to the central 0: and can then go to any of the 5 other Bs.

(1) In how many ways (without of course using the same cell twice) can his sister Anona spell out her name in her name-hex (Fig. 83)?

(2) In how many ways could Suivius (an ancient Roman who speaks through their mother, Mrs Nickels, when she is in a trance) spell his?

(3) And, finally, in how many ways could Mr Nickels—known to his friends as Semidimes?

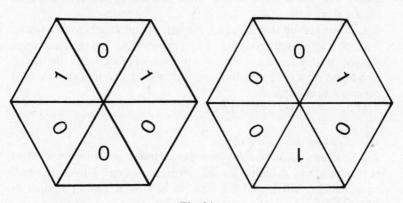

Fig. 84.

H5   Hexagon-cover

(1) In how many different ways can you colour the six triangles of a hexagon, using one of two colours, so that the resulting hexagons are distinguishable? (The two shown in Fig. 84 aren't distinguishable. Swivel one 120° and it is the same as the other).

(2) Now mark out the smallest possible pattern on a triangular grid, so that you can find each of those hexagon-patterns somewhere in it.

H6   Square dancing

At the Cricket Club dance, the first team (*minus* the opening batsmen) went onto the floor and stood smartly in a 'straight' square (Fig. 85A). When the applause died down, the opening batsmen and the

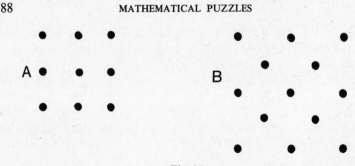

Fig. 85

scorer and the umpire joined them, and all 13 stood in a 'skew' square (Fig. 85B).

Encouraged by the clapping, a batch of supporters joined them: and this augmented group stood first in a straight and then in a skew square. With the rafters ringing, everyone present took the floor: and again they were able to—and did—stand first in a straight and then in a skew square.

How many were at the dance?

H7 Squares and killing them

Fig. 86 shows a square array of 16 pegs. How many different squares is it possible to draw with each corner at a peg? There are small squares, like ABFE and big squares like ACKI; straight squares like ADPM, and skew squares like BGJE.

(1) How many are there with this 4×4 array?

(2) How many are there with a 100×100 array?

If you remove pegs ABCHIJLMO, it isn't possible now to draw any squares at all. Is it necessary to remove as many as that, in order to 'kill' all the squares?

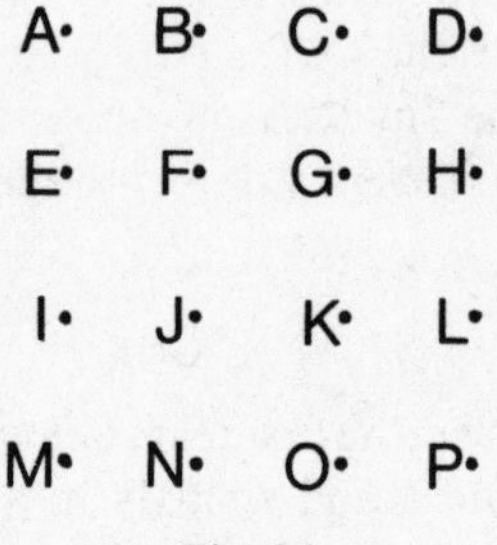

A· B· C· D·

E· F· G· H·

I· J· K· L·

M· N· O· P·

Fig. 86.

(3) How many pegs must you remove from a 4×4 array to 'kill' all possible squares?

(4) How many must you remove to kill all possible squares in a 100×100 array?

H8 Nino-squaro-recto

All that is known of the minisaurs' favourite game, nino-squaro-recto, is that it was played on a rectangular board, marked out in squares, and that the total number of rectangles on the board was nine times the total number of squares. Clearly it wasn't played on the 3×2 board shown in Fig. 87, because that has 8 squares (6 little, 2 big) and 18 rectangles (one of size 3×2, two of 3×1, two of 2×2, seven of 2×1, and six of 1×1); and 18 is not 9×8.

When I asked my three long-suffering nephews to work out how big the board was, each produced a different answer. Could they all have been right?

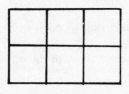

Fig. 87.

H9 Newt's and Coot's Grids

Fig. 88 shows a 2×1 grid. How many different straight lines of integral length could you choose from it? 7 of length 1 and 2 of length 2; 9 in all. Their total length is 11. So their average length is $1\frac{2}{9}$.

Newt and Coot each drew a grid, and calculated the average length of all the line-choices. The answers turned out to be the same

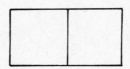

Fig. 88.

whole number. Newt's was a square grid. Coot's was double-square, i.e. twice as long as it was wide. Neither was as big as 30 either way.
How big were Newt's and Coot's grids?

H   Solutions

H1S   How few flyovers?

225 flyovers are needed to link 13 towns. The general 'flyover formula' for n towns is $[n(n-2)^2(n-4)]/64$, if n is even; $[(n-1)^2(n-3)^2]/64$, if n is odd.

One way to draw a motorway pattern with that number of flyovers is to place the towns in two concentric rings; $n/2$ in each, or $[n-1]/2$ in one and $[n+1]/2$ in the other; and then join

   (a) those in the inner ring to each other by the shortest route inside that ring:
   (b) those in the outer ring to each other by the shortest route outside that ring:
   (c) those in the inner to those in the outer by the shortest route across.

Fig. 89 shows that done for $n=8$, wherein towns are solid blobs and flyovers are 0s. Now it's easy to calculate that joining k points *inside* a ring costs $[k(k-1)(k-2)(k-3)]/24$ crossings: and so does joining k *outside* a ring (by turning the same pattern 'inside out'). And it works out that each of k in one ring can be joined to each of k in a concentric ring with $[k^2(k-1)(k-2)]/6$, and k to $(k+1)$ with $[k-1)^2k(k+1)]/6$. Adding these gives the general 'flyover' formula quoted at the beginning.

Linking n towns with the formula number of flyovers is certainly always possible; but I can't see how to prove that it isn't ever possible with fewer than that.

H2S   The committee function
8 before; 9 after.

If $A(k)$ is the number of ways a committee of k can 'agree or disagree', values of $A(k)$ up to $k=20$ are:

k	A(k)	Successive differences					
1	1						
2	2	1	2				
3	5	3	7	5			
4	15	10	27	20	15		
5	52	37	114	87	67	52	203
6	203	151	523	409	322	255	
		674					
7	877						
8	4,140		etc				
9	21,147						
10	115,975						
11	678,570						
12	4,213,597						
13	27,644,437						
14	190,899,322						
15	1,382,958,545						
16	10,480,142,147						
17	82,864,869,804						
18	682,076,806,159						
19	5,832,742,205,057						
20	51,724,158,235,372						

This series crops up in odd connexions; and it has some interesting features. First, the successive columns of 'differences' start with the terms of the series itself, which gives the series a certain uniqueness. Second, each term can be calculated by multiplying all the preceding terms by the appropriate binomial co-efficients: e.g., $877 = 1.203 + 6.52 + 15.15 + 20.5 + 15.2 + 6.1 + 1.1$ (note that we must take A(o) as 1). Third, the sum of inflnity of $n^k/n!$ involves the A(k) series. In fact,

$$\sum_{n=1}^{\infty} \left(\frac{n^k}{n!}\right) = A(k)e.$$

Fourthly, $x^4 = 1.x(x-1)(x-2)(x-3) + 6.x(x-1)(x-2) + 7.x(x-1) + 1.x$, and the sum of coefficients $1 + 6 + 7 + 1 = A(4)$. This is true generally, and leads to another way of evaluating the series. The

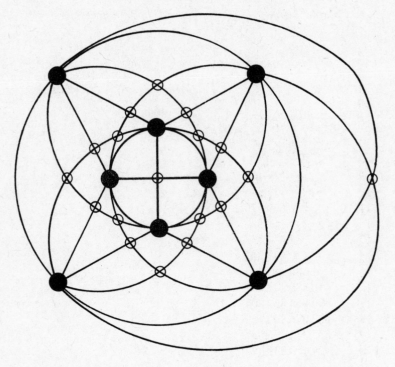

Fig. 89.

necessary table begins

K	Col 5	Col 4	Col 3	Col 2	Col 1	A(k)
1	.	.	.	.	1	1
2	.	.	.	1	1	2
3	.	.	1	3	1	5
4	.	1	6	7	1	15
5	1	10	25	15	1	52
	&c					

Each number in column c of the table is c times the number immediately above *plus* the number above-right; e.g., $25 = (3 \times 6) + 7$. And A(k) is the sum of all the numbers in row k.

But how find a *simple* formula for the series?

H3S   Name-squares
(1) Ada can score 36, as shown in Fig. 90A.

(2) 36, I think, is the best Anna can do, too: the number by each A in Fig. 90B shows the number of ANNA's starting there.

(3) Anona can score 256. (Fig. 90C).

(A)

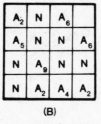

(B)

A	N	A	N	A
N	O	N	O	N
A	N	A	N	A
N	O	N	O	N
A	N	A	N	A

(C)

Fig. 90.

The easiest way to count, for a palindromic name with an *odd* number of letters, is to concentrate on the middle letter. With ANONA, for instance, from each O in the pattern shown you can spell ONA in 8 ways: so you can spell ANONA with that O in $8 \times 8 = 64$ ways. Since there are 4 Os, the total is 256.

H4S   Name-hexes
(1) 258

(2) 1,398

(3) 6,402

Fig. 91 shows part of the SEMIDIMES hex. The number by each letter is the number of routes to that cell from the central D. It is not difficult to see that, if you allow cells to be used twice, the total of routes for SEMID is $(1 + 4 + 6 + 4) \times 6 = 90$, so the total for SEMIDIMES is $90^2 = 8100$. Similarly, the total for EMIDIME (or SUIVIUS) is $(6 \times 7)^2 = 1764$; and that for MIDIM (or ANONA) is $(6 \times 3)^2 = 324$. In fact, if there are $2n + 1$ letters in the name, this gross total of routes $= 36(2^n - 1)^2$.

To find the answer we want, we must subtract from that total the number of routes which use one or more cells twice. Call this O (n). $O(1) = 6$ (the BOB case). $O(2)$ is the sum of

(a) routes where an I cell (using MIDIM) is the nearest to the

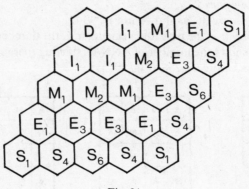

Fig. 91.

centre which is used twice: there are 6 Is, each of which has 3 links to an M, at each end of the word: $6 \times 3 \times 3 = 54$.

(b) routes where an M is the innermost duplicated cell: only with the M2s—not the M1s—can a MIDIM have different Is but the same M: there are 6 M2s, and each such MIDIM can be traversed in 2 directions: $6 \times 2 = 12$.

Hence $O(2) = 54 + 12 = 66$. By similar reasoning,

$$O(3) = 294 + 48 + 24 = 366$$
$$O(4) = 1350 + 192 + 96 + 60 = 1,698$$
$$O(5) = 5,766 + 768 + 384 + 240 + 168 = 7,326$$

Hence the answers given, and the result for Sexipopixes—27,270. It would be nice to find an explicit formula for these results, or for $O(n)$.

H5S   Hexagon-cover

(1) There are 14 different hexagon-patterns.

(2) The smallest pattern containing n different hexagons turns out to contain $2n + 1 + \sqrt{12n - 3}$ (or, if that is not integral, the next integer above that) triangles. So, for $n = 14$, we need at least a 42-triangle pattern, such as the one shown in Fig. 92. Trial quite easily shows it is possible to colour it to exhibit all 14 varieties, as shown.

H6S   Square dancing

841. A skew square consists of two nested straight squares, whose

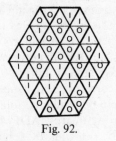

Fig. 92.

edges differ by 1. So we need to solve

$$N = y^2 = x^2 + (x+1)^2,$$

i.e.

$$2x^2 + 2x - (y^2 - 1) = 0,$$

whence

$$x = \frac{\sqrt{2y^2 - 1} - 1}{2}, \quad \text{or} \quad (1) \ x = \frac{z-1}{2},$$

$$\text{and} \quad (2) \ z^2 - 2y^2 = -1.$$

Now (2) is Pell's equation. One way to solve it is to say that $z = 2y^2$ nearly, i.e. $(z/y) = \sqrt{2}$ nearly. Answers occur among fractions which are 'best approximations' to $\sqrt{2}$, i.e. $\frac{1}{1}, \frac{3}{2}, \frac{7}{5}, \frac{17}{12}, \frac{41}{29}, \frac{99}{70}, \frac{239}{169}, \ldots$ Alternate values give us what we want,

z	y	x	$x+1$	N
1	1	0	1	1
7	5	3	4	25
41	29	20	21	841
239	169	119	120	28,561

I know no dance floor which would take 28,561. So there were 25 when the first batch of supporters joined in. And there were 841 in all at the dance.

H7S Squares and killing them
(1) 20
(2) 8,332,500
(3) 6 (say ACFHJO)
(4) 4,950 (I believe)

To solve (2), we need to develop a general formula for an N^2 array. One way to do this is to classify the squares according to the smallest orthogonal square of pegs which will contain them. Then each smallest peg-square, such as ABFE, corresponds to 1 square: there are $(n-1)^2$ of these peg-squares: total $1.(n-1)^2$. Each next smallest peg-square (e.g. ACKI), corresponds to 2 squares (ACKI and BGJE): there are $(n-2)^2$ of these peg-squares: total $2.(n-2)^2$. Peg-square ADPM corresponds to 3 squares (ADPM, BHOI, CLNE): and there are $(n-3)^2$ of these peg-squares: total $3(n-3)^2$. And so on, giving a grand total of $(n-1)^2+2(n-2)^2+3(n-3)^2\ldots$

$$\text{i.e.} \quad \sum_{k=1}^{n-1} k(n-k)^2, \quad \text{which} = \frac{n^2(n^2-1)}{12}.$$

To solve (4) we also need a formula. The one which works for small n, and on which the answer given for (4) is based, is that you need to remove $[n(n-1)]/2$ in order to kill all possible squares. But I don't see how to prove it.

H8S   Nino-squaro-recto

Yes. Boards of 6×15, 8×14, and 11×12, all satisfy the condition.

The general formula for the number of *squares* in a $k \times n$ rectangle

$$(k < n) \quad \text{is} \quad \frac{k(k+1)(3n-k+1)}{6}.$$

The number of *rectangles is*

$$\frac{k(k+1)n(n+1)}{4}.$$

If the latter is to be a multiple M of the former, we have to solve

$$M = \frac{3n(n+1)}{2(3n-k+1)}$$

Here $M = 9$. So $18(3n-k+1) = 3n(n+1)$,

$$\text{i.e.} \quad k = \frac{n(17-n)}{6} + 1.$$

Trial shows there are three solutions with $n > k$,

Sides	Squares	Rectangles
$n=15, k=6$	280	2520
$n=14, k=8$	420	3780
$n=12, k=11$	572	5148

Note: solutions for other small M are:

M	Sides	Squares	Rectangles
3	$n=4,\ \ k=3$	20	60
6	$n=8,\ \ k=7$	168	1,008
10	$n=15, k=10$	660	6,600
12	$n=16, k=15$	1360	16,320

H9S   Newt's and Coot's Grids
Newt's was 10×10, Coot's 6×12. Average line-length in both was 4.
The average line-length in an $a \times b$ grid is

$$\frac{a(a+2)+b(b+2)}{3(a+b)}$$

So if $a=b$ (as with Newt), the average is $(a+2)/3$; while if $2a=b$ (as with Coot), the average is $(5a+6)/9$.
   The lowest solutions are:

> Newt 10×10:   Coot 6×12
> Newt 25×25:   Coot 15×30

The latter is too big, so the first is the answer.
Note: the number of different lines you can choose from an $a \times b$ grid is

$$\frac{(a+1)(b+1)(a+b)}{2}.$$

I Ominos and Other Animals

11 Omino-spoiling

Suppose I have a set of the 12 different pentominos (Fig. 93A) and put them one by one onto a 5×5 grid. I always put them 'properly', i.e. edges precisely on grid-lines. If I start by placing the V and W as in Fig. 93B, I cannot place any of the other 10 pentominos. I have 'spoiled' 5^2 with 2 pentominos: and obviously 5^2 can't be spoiled with fewer.

(1) With how few pentominos can you spoil a grid of 6^2? Of 7^2?—and so on up to 13^2?

(2) And with how few of the 5 tetrominos (combinations of 4 unit squares) can you spoil a 5^2 grid? A 6^2 grid? (You can of course place ominos either side up).

12 Omino holes

(1) How many separate 'holes' can you enclose with n out of the 12 different pentominos shown in Fig. 93A? Each hole is to be one unit square. Try with 2, 3, 4 . . . 12 pentominoes.

(2) How *large* a hole can you enclose with all 12 pentominos?
Rules (a) no hole may be in contact with another hole, or with the outside world, even at a corner; (b) ominos must be placed 'properly', i.e. edges on lines of a unit-square grid; (c) ominos may be placed either side up.

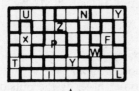

A

B

Fig. 93.

13   Holes with iamonds

(1) How many separate holes can you enclose with the set of 4 pentiamonds shown in Fig. 94A?

(2) How many with n out of the 12 different hexiamonds shown in Fig. 94B? Try with 2, 3, 4 ... 12.

(3) How *big* a hole can you enclose with the 4 pentiamonds?

(4) How big a hole with the 12 hexiamonds?

Rules   (a) No hole may touch another hole, or the outside world, even at a corner:

      (b) iamonds must be placed 'properly'—i.e. edges on lines of an equilateral-triangle-grid:

      (c) iamonds may be placed either side up.

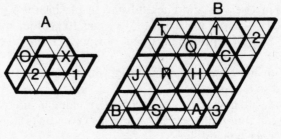

Fig. 94.

14   Quadrabolo holes

(1) How many separate holes can you enclose with n out of the 14 different quadrabolos shown in Fig. 95? Each hole is to be one small right-angled isosceles triangle.

Fig. 95.

Try with $n = 3, 4, 5 \ldots 14$ quadrabolos.

(2) How *large* a hole can you enclose with all 14?

Rules   (a) no hole may touch another hole, or the outside world, even at a corner:

(b) bolos must be placed 'properly'—i.e. edges on lines or square-diagonals of a square-grid:

(c) bolos may be placed either side up.

15 Fitting ominos in

Asked to design the smallest pattern of squares into which you could fit any one of the 2 tromino patterns (marked '3' in Fig. 96A), Hermione rightly specified the patterns in Fig. 96B and C.

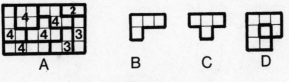

Fig. 96.

(1) What is the smallest pattern which will accommodate any of the 5 tetrominos (marked '4' in Fig. 96A).

(2) And what for any of the 12 pentominos (shown in Fig. 93A)?

(3) And what for any of the 35 hexominos?

(4) ... for the 108 heptominos (including the awkward customer shown in Fig. 96D)?

16 Awkward omino maps

The map in Fig. 97 has 7 districts, and also 2 lakes (hatched). The

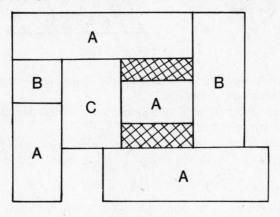

Fig. 97.

map can be properly coloured with 3 colours, A, B and C, as shown—
'properly' in that any two districts with a bit of edge in common have
different colours.

An 'awkward' map is one that cannot be properly coloured with
3 colours.

What is the smallest awkward map—smallest in area, that is,
including any lakes—whose districts are all

    (1) 4-minos?
    (2) 3-minos?
    (3) 2-minos?
    (4) 1-minos?
    (5) 5-minos?

17   Convos and trimps

An n-convo is a convex arrangement of n equal isosceles right-angled
triangles, i.e. a convex n-bolo. An n-trimp is a convex arrangement of
n equal equilateral triangles, i.e. a convex n-iamond. Fig. 98 shows
all the different convos from 1 to 3 put together to form one 13-
convo, and all the different trimps from 1 to 7 put together to form a
45-trimp.

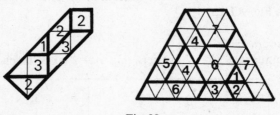

Fig. 98.

Two convos (or trimps) which are the same shape when turned
over do not count as different.

(1) How many different n-convos and n-trimps are there, for
values of n up to 30?

(2) What is the smallest value of n for which you can arrange all
the trimps from 1 up to n in an equilateral triangle?

18   Tetracubes

Fig. 99 shows a full set of 8 tetracubes—arrangements of 4 cubes
joined face to face. You can put the set together to form various

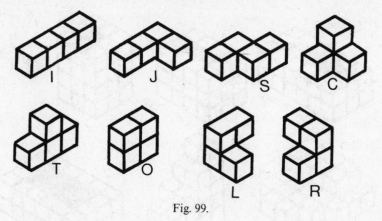

Fig. 99.

patterns in three dimensions. Try forming the following:

(1) Solids $8 \times 2 \times 2$

$\qquad 4 \times 4 \times 2$

(2) A replica of each individual tetracube, twice as big in each dimension,

(3) Patterns A–K in Fig. 100.

All but one of these are possible.

I Solutions

IIS Omino-spoiling

Fig. 101 shows the best solutions I have found. Few of the patterns are unique.

I2S Omino holes

(1) Pentominos	2	3	4	5	6	7	8	9	10	11	12
Holes		1	2	3	4	6	7	8	9	11	12	13

Fig. 102A shows solutions up to 9 pentominos, if you add them in order I U Y X L Z N W V. Fig. 102B solves 10 and (with the F-pentomino) 11. Fig. 102C solves all 12. I think a pattern for 9 pentominies enclosing 10 holes may be possible, but I haven't found one.

(2) 128 squares (Fig. 103) is the largest hole I have been able to enclose.

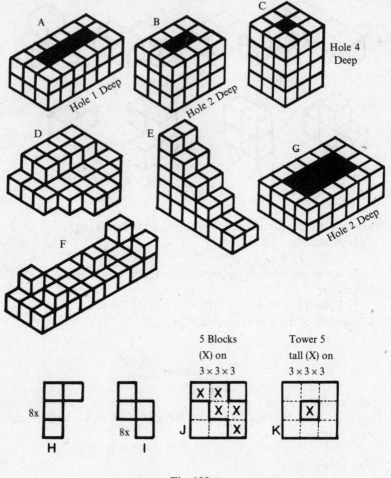

Fig. 100.

13S   Holes with iamonds

(1) Two (Fig. 104a).

(2)
Hexiamonds	1	2	3	4	5	6	7	8	9	10	11	12
Holes	0	1	1	2	3	4	5	6	7	7	8	9

Fig. 104B, placing pieces in the order 1, 2, 3 etc., shows how. I should

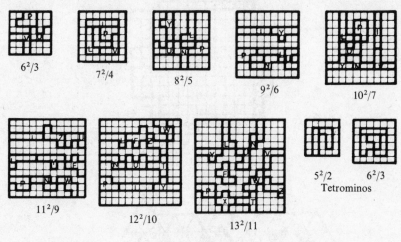

$6^2/3$ $7^2/4$ $8^2/5$ $9^2/6$ $10^2/7$

$11^2/9$ $12^2/10$ $13^2/11$

$5^2/2$ $6^2/3$
Tetrominos

Fig. 101.

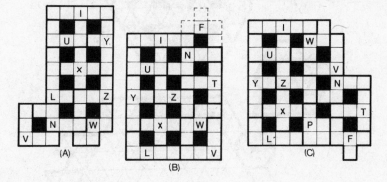

(A)   (B)   (C)

Fig. 102.

like to manage one more hole with 10, 11, and 12: but how?
(3) 5 (Fig. 104C).   (4) 113 (Fig. 104D) is my best.

I4S   Quadrabolo holes

(1) Quadrabolos	3/4	5	6/7	8/9	10/11	12/13	14
Holes		1	2	3	4	5	6	7

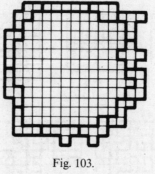

Fig. 103.

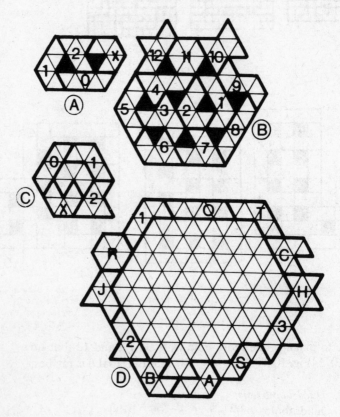

Fig. 104.

Fig. 105A shows 3 holes with 6 bolos, and also yields patterns for 2 holes with 5 and 1 hole with 3.

Fig. 105B without BHMN, shows 4 holes with 8; without MN, 5

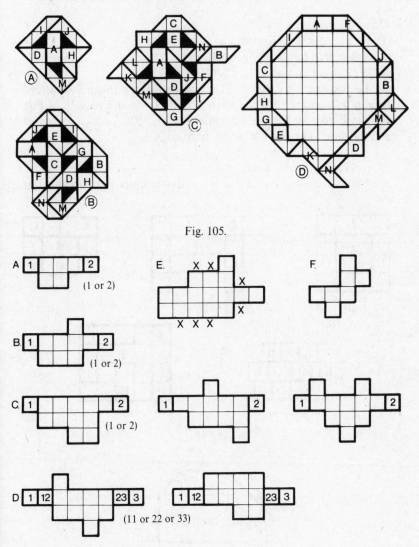

Fig. 105.

Fig. 106.

holes with 10; and, as it stands, 6 with 12.

Fig. 105C shows 7 holes with all 14. I wonder if 8 holes is feasible?

(2) $43\frac{1}{2}$ squares (87 units) is the best I have contrived (Fig. 105D).

15S   Fitting ominos in

(1) You need a 6-square pattern (Fig. 106A).

(2) A 9-square pattern (Fig. 106B).

(3) A 12-square pattern (Fig. 106C or D).

(4) The sequence 1, 2, 4, 6, 9, 12 strongly suggest a 16-square pattern here, but I cannot do better than 17. The 16-square pattern in Fig. 106E will take all but the heptomino in Fig. 106F, which forces an extra square at one of the points marked X.

16S   Awkward omino maps

See Fig. 107. To make sure that (4) needs 4 colours, note that the top

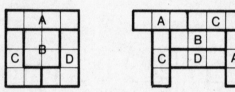

(1) 16   (2) 18   (3) 12

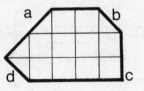

(4) Just over 11   (5) 20

Fig. 107.

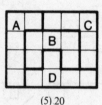

Fig. 108.

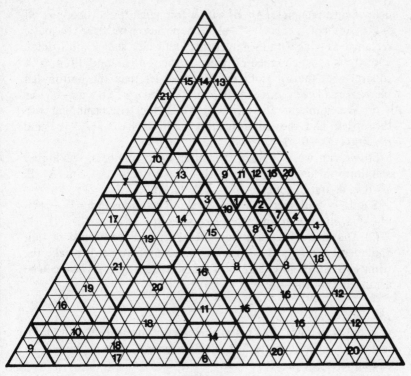

Fig. 109.

A forces A on left and right and that these would force two touching A's at the middle bottom.

I7S Convos and trimps

n	1	2	3	4	5	6	7	8	9	10	11	12	13	14	15
Convos	1	3	2	6	3	7	5	11	5	10	7	14	7	16	11
Trimps	1	1	1	2	1	2	2	3	2	2	2	3	2	3	3

n	16	17	18	19	20	21	22	23	24	25	26	27	28	29	30
Convos	20	9	17	13	22	12	25	18	27	14	24	21	31	21	36
Trimps	5	2	3	3	4	2	4	4	6	3	3	4	5	2	5

(1) One way of evaluating convos is to use the fact that each can be

defined as a rectangle (A × B) with a triangular 'piece' chopped off 0, 1, 2, 3, or 4 of its corners. To avoid duplication, we regard the basic rectangle as lying with its longer side (if it isn't a square) horizontal, with the biggest chopped-off piece at the top left, and (if it is a square) with the top-right piece no smaller than the bottom-left piece. We call each convo 'A, B; a, b, c, d,' and $n = 2AB - a^2 - b^2 - c^2 - d^2$. We enumerate the pieces a, b, c, d clockwise beginning with the top left. Thus the convo illustrated in Fig. 108 is '5,3; 2, 1, 0, 1' and $n = 30 - 4 - 1 - 0 - 1 = 24$.

So we can work if we wish entirely with numbers, seeking all solutions in integers of $n = 2AB - a^2 - b^2 - c^2 - d^2$, with $A \geqslant B$, $a \geqslant b$, c, d, and of course $A \geqslant a + b, B \geqslant a + d$.

Similarly, with trimps, we can seek all solutions of $n = P^2 - q^2 - r^2 - s^2$, with $q \geqslant r \geqslant s$ and $P \geqslant q + r + s, P > q$.

(2) The area of a large equilateral triangle composed of unit equilateral triangles is a square number, and we need to take the trimps up to 21 before their total area is a square (625). These can be arranged as in Fig. 109.

18S   Tetracubes

G is the only impossible pattern. However you place C and O, for instance, you split the wall into two segments, each containing an *odd* number of cubes.

J Diophantine

J1 Ages and squares

(1) Bill has 2 children; Cecil has 3. The sum of the children's ages is the same for each family; so is the sum of the squares of their ages. What is the youngest set of ages that makes this possible?

(2) Don has 2 children, Eric 3, and Fred 3. Again, the age-sums are the same, and the sums of the squares of the ages are the same, in all the families. What is the youngest set of ages? (In each case, all ages are whole numbers and all are different.)

J2 Ages, twins and squares

(1) Annie has 2 children, twins; Bonnie has 3 children. The sum of the children's ages is the same for each family; so is the sum of the squares of their ages. What is the largest number of the 5 children that can be teenagers?

(2) Connie has 2 children; Denise has 3 (a pair of twins and one other). Again, the age-sum is the same for each family; and so is the sum of squares of ages. How many of these 5 can be teenagers?

(3) Ena and Fiona have the squarest families in the whole neighbourhood. Ena has 3 children, Fiona 3. In each family, the children's ages add up to the same *square* number; and the squares of the children's ages add up to the same number, and that too is *square*. How old are the children? (In each case, all ages are whole numbers).

J3 Ages and cubes

(1) Gerry has 2 children: Hywel has 3. The sum of the children's ages is the same for each family: so is the sum of the cubes of their ages. What is the youngest set of ages that makes this possible?

(2) Ian has 2 children: Jack 3: Ken 2; Leonard 3: The age-sums of all four lots of children are the same. As for sums of cubes of ages,

111

Ian's and Jack's children are equal, and so are Ken's and Leonard's. One of Leonard's is the oldest of the 10. What is the youngest possible set of ages? (In each case, all ages are whole numbers and all are different).

J4   Ages and products

(1) Maurice and Nick have 3 children each. The sum of the children's ages is the same for each family: so is the product of the ages. What is the youngest set of ages that makes this possible?

(2) Olga, Pam, and Queenie have 3 children each. For each family, the age-sum and the age-product are the same. What is the youngest possible set of ages? (In each case, all ages are whole numbers and all are different).

J5   Ages and reciprocals

Ronny has two sons: Sonny has 3. The sum of the sons' ages is the same for each family. So is the sum of the reciprocals of the sons' ages. How old are the sons?

If you suggest that Sonny's are 65, 104 and 120, and Ronny's 34 and 255, I will agree that each family adds up to 289, and that the reciprocals add up to $\frac{1}{30}$ in each case; but Sonny and Ronny aren't as old as that would imply. Each is under a hundred. (All ages are whole numbers and all are different).

J6   Treasure in a field

X in Fig. 110 marks the spot where the treasure is buried in this

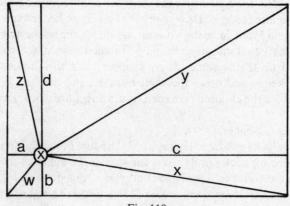

Fig. 110.

rectangular field. The 8 distances from X to all four sides and to all four corners are all different, and all are whole numbers of yards. Actually, the distances are $a = 15$, $b = 20$, $c = 48$, $d = 36$, $w = 25$, $x = 52$, $y = 60$, $z = 39$.

That was the best Uncle Geoffrey, who buried the treasure, could do in a field only 63 yards × 56 yards. But he thinks the arrangement less than ideal, because the proportions a/b and d/c are the same, and so are the proportions b/c and a/d. So he is going to buy a new field—bigger, but its perimeter is still under 1000 yards—and rebury the treasure in it in such a way that the 8 distances to sides and corners are all different whole numbers of yards *and* the proportions of all four rectangles (a/b, b/c, c/d, d/a) are quite different.

Can you draw his new field, with X marking where the treasure will be?

J7 Py-bricks
Fig. 111 shows the Py-brick Elizabeth Anne designed. A Py-Brick

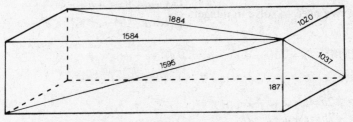

Fig. 111.

is of course a rectangular parallelepiped, with all its edges and the diagonals of all its faces each an exact whole number of millimetres.

She would like a smaller Py-Brick—one, that is, with a smaller edge-sum. Her design, as you see, sums to $187 + 1020 + 1584 = 2791$. Can you help her?

J8 Triangle areas and perimeters
(1) How many more triangles can you find which, like the first one shown in Fig. 112, have whole-number sides whose length adds up to the area of the triangle?
(2) And how many more which, like the second one shown, have whole-number sides differing by 1, and have a whole-number area? (Note: in neither case do the triangles have to be right-angled).

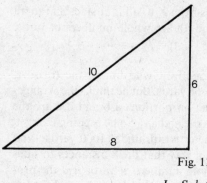
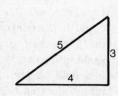

Fig. 112.

J   Solutions

J1S   Ages and squares
(1) Bill's are 4 and 5: Cecil's 1, 2, and 6.
(2) Don's are 11 and 13: the others 1, 8, and 15 and 3, 5, and 16.

1. We need to solve in integers $\begin{cases} M = a+b = c+d+e \\ N = a^2+b^2 = c^2+d^2+e^2. \end{cases}$

2. $M^2 - N = 2a(c+d+e-a) = 2(cd+ce+de).$

$$\therefore\ a^2 - (c+d+e)a + (cd+ce+de) = 0$$

i.e., $\left.\begin{aligned} 2a &= c+d+e+Z \\ 2b &= c+d+e-Z \\ Z^2 &= (c+d+e)^2 - 4(cd+ce+de) \end{aligned}\right\}$

$4cd = (c+d-e)^2 - Z^2 : \therefore\ cd = \left(\dfrac{c+d-e+Z}{2}\right)\left(\dfrac{c+d-e-Z}{2}\right) = pqrs,$
say.

3. $\begin{aligned} c &= pq & c+d-e+Z &= 2pr \\ d &= rs & c+d-e-Z &= 2qs \\ c+d-e &= pr+qs & e &= pq+rs-pr-qs = (p-s)(q-r) \\ Z &= pr-qs & a &= pr+e = pq+rs-qs \\ & & b &= qs+e = pq+rs-pr \end{aligned}$

4. Now put $(p-s) = t$; $(q-r) = u$, and we get

$$d = rs:\quad e = tu:\quad c = (s+t)(r+u):\quad b = c-rt:\quad a = c-su.$$

5. To avoid cancellable factors, and mere rearrangements, we stipulate

r, u coprime
s, t coprime
$r < u$, or $r = u$ and $s \leqslant t$
$r < s$, or $r = s$ and $u \leqslant t$
$r \leqslant t$.

6. We can now tabulate solutions for as long as we are interested. The first solution with all different a, b, c, d, e, will solve problem (1). For problem (2), we need to go a little further.

r	u	s	t	b	a	c	d	e	M	N
1	1	1	1	3	3	4	1	1	6	18
1	1	1	2	4	5	6	1	2	9	41
1	1	1	3	5	7	8	1	3	12	74
1	2	2	1	5	8	9	2	2	13	89
1	2	1	2	7	7	9	1	4	14	98
1	1	2	3	7	8	10	2	3	15	113
1	1	1	4	6	9	10	1	4	15	117
1	2	3	1	6	11	12	2	3	17	157
1	1	1	5	7	11	12	1	5	18	170
1	2	1	3	9	10	12	1	6	19	181
1	1	3	4	10	11	14	3	4	21	221
1	1	2	5	9	12	14	2	5	21	225
1	1	1	6	8	13	14	1	6	21	233
1	2	4	1	7	14	15	2	4	21	245
1	2	3	2	9	13	15	3	4	22	250
1	3	3	1	7	15	16	3	3	22	274
1	2	2	3	11	12	15	2	6	23	265
1	2	1	4	11	13	15	1	8	24	290 ⎫
1	1	3	5	11	13	16	3	5	24	290 ⎭
1	1	1	7	9	15	16	1	7	24	306

and so on until

1	1	1	9	11	19	20	1	9	30	482 ⎫
1	2	5	2	11	19	21	4	5	30	482 ⎭

7. The general method of which this is an example can be used in quite a number of Diophantine ('solve in integers') problems.

J2S Ages, twins and squares

(1) 3 teenagers if A's are 13 and 13, B's 1, 9, and 16.

(2) 2 teenagers if C's are 7 and 15, D's 3, 3, and 16.

(3) E's are 21 and 28; F's are 1, 18, and 30—giving a sum of 49 and a square-sum of $35^2 = 1225$.

The general solution to problem (1) is
$(a^2 + a + 1), (a^2 + a + 1); 1, a^2, (a + 1)^2$, giving

$$3, 3; 1, 1, 4$$
$$7, 7; 1, 4, 9$$
$$13, 13; 1, 9, 16$$
$$21, 21; 1, 16, 25; \text{etc.}$$

The general solution to (2) is
$a(a + 2), 2a + 1; a, a, (a + 1)^2$, giving

$$3, 3; 1, 1, 4$$
$$8, 5; 2, 2, 9$$
$$15, 7; 3, 3, 16$$

$$\dots\dots\dots$$

$$195, 27; 13, 13, 196$$

(which is hardly realistic).

As to (3), the system

$$a + b = c + d + e = X^2$$
$$a^2 + b^2 = c^2 + d^2 + e^2 = Y^2$$

is not soluble with smaller X and Y than 7 and 35. I do not know if solutions with larger X and Y are possible.

J3S Ages and cubes

(1) Gerry's are 7 and 8: Hywel's 1, 5, and 9.

(2) Ian's 32 and 45: Jack's 11, 17, and 49. Ken's 23 and 54: Leonard's 8, 14, and 55.

A method similar to that explained in the Solution to 'Ages and squares' leads me to the formulae:

$$
\left.
\begin{aligned}
a &= ij(2kn - in - jk) \\
b &= k(n - j)(kn - in - ij) \\
c &= n(2ijk - i^2j - k^2n) \\
d &= (k - i)(kn^2 - ij^2) \\
e &= (kn - ij)(kn + ij - jk)
\end{aligned}
\right\}
\quad
\begin{aligned}
a + b &= c + d + e = M, \\
a^3 + b^3 &= c^3 + d^3 + e^3 = N.
\end{aligned}
$$

but without entire confidence that it gives all solutions.

From this we get the solutions given above. For $M = 77$, there are three solutions

$$32, 45: 11, 17, 49$$
$$23, 54: \ 8, 14, 55$$
$$27, 50: 11, 14, 52,$$

but only the first two have no number in common.

J4S   Ages and products
(1) $12 + 4 + 3 = 9 + 8 + 2 = 19$

$\quad 12 \times 4 \times 3 = 9 \times 8 \times 2 = 144$

(2) $24 + 10 + 5 = 20 + 15 + 4 = 25 + 8 + 6 = 39$

$\quad 24 \times 10 \times 5 = 20 \times 15 \times 4 = 25 \times 8 \times 6 = 1200$

J5S   Ages and reciprocals
Ronny's are 45 and 5: Sonny's are 24, 18, and 8. In each case the sum is 50 and the sum of reciprocals is $\frac{2}{9}$.

We have to solve $a + b + c = d + e$

$$\text{and} \quad \frac{1}{a} + \frac{1}{b} + \frac{1}{c} = \frac{1}{d} + \frac{1}{e}$$

Putting $X = a + b + c$

$\quad Y = ab + bc + ca$

$\quad Z = abc$, it is not hard to reach the formula for d and e

$$\frac{1}{2}\left[X \pm \sqrt{X^2 - 4\,\frac{XZ}{Y}} \right].$$

I find this hard to manage. But if we put $a = x(x+y)$, $b = y(x+y)$, $c = pxy$, we get a more reasonable formula, and if we put $p = 3$, it simplifies further to

$$\tfrac{1}{2}\left[(x+y)^2 + 3xy \pm (x+y)\sqrt{(x+y)^2 + 3xy}\right].$$

$x = 5$, $y = 8$ in that equation gives the solution quoted in the question, and $x = 1$, $y = 3$ gives a solution which can be doubled to give the answer quoted here. But the simplified equations don't of course give all the possible solutions, and an answer in smaller numbers may well be possible.

J6S   Treasure in a field

The new field has a perimeter of 976 yards, as shown in Fig. 113.

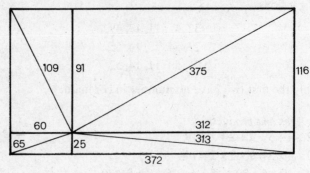

Fig. 113.

Clearly we need

$$\frac{a}{b}=\frac{A1}{A2} : \frac{b}{c}=\frac{B1}{B2} : \frac{c}{d}=\frac{C1}{C2} : \text{ and } \frac{d}{a}=\frac{D1}{D2}, \text{ and } \frac{abcd}{bcda}=\frac{A1 \times B1 \times C1 \times D1}{A2 \times B2 \times C2 \times D2}=1$$

where A1/A2 etc. are 'Pythagorean fractions' ('pyfracs' for short), in which numerator and denominator are feasible non-hypotenuse sides of an integral right-angled triangle, in lowest terms. Small pyfracs are $\frac{3}{4}, \frac{5}{12}, \frac{7}{24}, \frac{11}{60}, \frac{39}{80}, \frac{65}{72}, \frac{60}{91}$, etc. and of course their reciprocals $\frac{4}{3}, \frac{12}{5}$, etc.

Solutions are quite easy to find by putting

$$\frac{A1}{A2}=\frac{B2}{B1} \text{ and } \frac{C1}{C2}=\frac{D2}{D1}.$$

For instance $\frac{3}{4} \times \frac{4}{3} \times \frac{5}{12} \times \frac{12}{5}=1$ yields Uncle Geoffrey's original field. But to get all four rectangles with different proportions, we need four different pyfracs. Trial and error shows 3 foursomes of smallish pyfracs which multiply out to 1:

$$\frac{5}{12} \times \frac{40}{9} \times \frac{39}{80} \times \frac{72}{65}=1$$

$$\frac{3}{4} \times \frac{7}{24} \times \frac{60}{11} \times \frac{88}{105}=1$$

$$\frac{5}{12} \times \frac{7}{24} \times \frac{60}{91} \times \frac{312}{25}=1$$

Each of these can be arranged to produce 6 different solutions. It is one arrangement of the 3rd one that yields, after cancellation of factors, the solution drawn out above, which seems to be the smallest (but I can't prove it is).

J7S Py-bricks

$44 \times 117 \times 240$ (diagonals 125, 267, 244), with edge-sum of 401, is the smallest I can find.

If the brick is $a \times b \times c$, we clearly need

$$(1) \quad \frac{a}{b} = X : \frac{b}{c} = Y : \frac{c}{a} = Z$$

and $$(2)\ X.\ Y.\ Z. = 1.$$

where X, Y and Z are what I call 'py-fracs'—fractions of the form $(a^2 - b^2)/2ab$ or $2ab/(a^2 - b^2)$ (with a, b coprime and of opposite parity), which are feasible proportions between the two non-hypotenuse sides of an integral right-angled triangle.

Now EA's design depends on $\frac{60}{11} \times \frac{17}{144} \times \frac{132}{85} = 1$. But the same py-fracs in the order $\frac{60}{11} \times \frac{132}{85} \times \frac{17}{144}$ yield a smaller brick than hers: $85 \times 132 \times 720$, summing to 937.

Taking py-fracs with numerator and denominator under 1000, I have found 8 solutions to equation (2), yielding 16 Py-bricks

py-fracs	Py-bricks					
$\frac{60}{11}, \frac{44}{117}, \frac{39}{80}$	44,	117,	240	429,	880,	2340
$\frac{20}{21}, \frac{252}{275}, \frac{55}{48}$	240,	252,	275	1008,	1100,	1155
$\frac{60}{11}, \frac{132}{85}, \frac{17}{144}$	85,	132,	720	187,	1020,	1584
$\frac{24}{7}, \frac{20}{99}, \frac{231}{160}$	160,	231,	792	140,	480,	693
$\frac{52}{165}, \frac{176}{57}, \frac{855}{832}$	832,	855,	2640	2964,	9152,	9405
$\frac{52}{165}, \frac{748}{195}, \frac{225}{272}$	780,	2475,	2992	2925,	3536,	11220
$\frac{104}{153}, \frac{748}{195}, \frac{135}{352}$	1755,	4576,	6732	1560,	2295,	5984
$\frac{252}{275}, \frac{44}{483}, \frac{575}{48}$	528,	5796,	6325	1008,	1100,	12075

If only one of these had $a^2 + b^2 + c^2 = D^2$! If so, the brick's diameter (corner to corner through the centre) would be a whole number too, and we should have a truly Integral Brick. Another and perhaps simpler trial-and-error approach to that problem is to find the various solutions for successive c in the system

$$c^2 = y^2 - b^2 = z^2 - a^2 = D^2 - x^2,$$

and hope to find $a^2 + b^2 = x^2$. But I haven't found an Integral brick, either way, yet.

J8S Triangle areas and perimeters
(1) Four others:

> Sides 5, 12, 13: Area and perimeter 30.
> Sides 6, 25, 29: A and P 60
> Sides 7, 15, 20: A and P 42
> Sides 9, 10, 17: A and P 36.

(2) An infinite number.
(1) We need, if the sides are *abc*,

$$a+b+c=\tfrac{1}{4}\sqrt{(a+b+c)(a+b-c)(a-b+c)(-a+b+c)},$$

i.e.

$$16(a+b+c)=(a+b-c)(a-b+c)(-a+b+c).$$

The 3 factors on the right are clearly all even or all odd: and they must be all even. Now put $(a+b-c)=2k$, $c=a+b-2k$, and we get

$$b=\frac{(k^2+4)(a-k)}{ka-(k^2+4)}.$$

Testing this by putting $k = 1$, 2 etc. successively quickly yields the 5 solutions. With $k=1$, we get the last 3 listed above, all obtuse-angled. With $k=2$, we get the two right-angled solutions. And higher k reveal no new solutions.

(2) If the sides are $(a-1)$, a, $(a+1)$, the formula for the area of a triangle boils down to $\tfrac{1}{4}\sqrt{3a^4 - 12a^2}$ or $(a/4)\sqrt{3(a^2-4)}$.

So we need $a^2 - 4 = 3b^2$, and A $= 3ab/4$.

Now $a^2 - 3b^2 = 4$ is only possible if a and b are both even (an odd square leaves a remainder 1 when divided by 4). So we can say $a=2c$, $b=2d$, A $= 3cd$, and $c^2 - 3d^2 = 1$. We can get all integral solutions to this from alternate partial fractions in the continued fraction for $\sqrt{3}$, thus

	c	d	$c^2 - 3d^2$
Continued	0	1	−2
Fraction	1	0	1
1	1	1	−2
1	2	1	1
2	5	3	−2
1	7	4	1
2	19	11	−2
1	26	15	1
2	71	41	−2
1	97	56	1
2	265	153	−2
1	362	209	1

So there is an endless series of solutions, of which the smallest are:

	Sides		*Area*
1,	2,	3	0
3,	4,	5	6
13,	14,	15	84
51,	52,	53	1,170
193,	194,	195	16,296

K   Arranging Numbers

K1   Two stamps

As the Postmaster-Supreme of newly independent Milharton, I have decided that postage will be a penny an ounce, that no more than 2 stamps may be put on a pracel, and that stamps of n different denominations will be available. I have not yet decided what number n is nor what denominations the different stamps will be.

Last week I asked young Fran to advise me; for $n = 2, 3, 4$ etc., what is the best set of denominations to choose, so that customers will be able to stamp parcels of every exact number of ounces up to the highest possible figure?

Today she said 'I am torn between two systems. Under System O, you have stamps of the first $(n-1)$ odd numbers, and also a $(2n-2)$ stamp. For instance, if $n = 8$, you have stamps of 1, 3, 5, 7, 9, 11, 13, 14; and you can stamp parcels up to 28 oz.

'Under System T, the first half of the stamp set is 1, 2, 3, ... up to $n/2$ (or up to $(n+1)/2$ if n is odd); thereafter they rise in value by jumps of $[(n/2)+1]$ (or $[((n+1)/2)+1]$ if n is odd). Thus, if $n = 8$, you have stamps of 1, 2, 3, 4, 9, 14, 19, 24; and you can also stamp parcels up to 28 oz.'

Should I take her advice, and if so should I use System O or System T, for values of n up to 20?

K2   Two-coin currencies

(1) In the Giggleswick of my imagination—as also in Biggleswade and Wigglesworth—there are coins of only two denominations in use: everything costs an exact number of thoils. Since neither coin is a one-thoil piece, it is not possible to pay for something costing one thoil without getting change. In fact there are exactly 9 different

sums (including one thoil) which you can't pay exactly without getting change. What is the largest of these sums?

(2) What are the two coins in use in each of the three towns, assuming that the bigger coin is biggest in Biggleswade and the smaller coin biggest in Giggleswick?

K3   Permo-bracelets

Fig. 114 shows a bracelet of beads, some red (0) and some blue (1). It

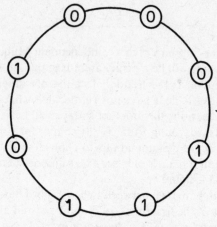

Fig. 114.

is a 'permo-bracelet'—actually a '3-permo-bracelet'—because you can find on it each of the possible permutations of 3 beads of either colour, 000, 001, 010, 011, 100 and so on. You can find each somewhere on the bracelet, in the right sequence; and you can go clockwise or anti-clockwise. Can you design, with the fewest possible beads,

(1) a 4-permo-bracelet?
(2) a 5-permo-bracelet?

K4

A Mixed Bag contains n different counters, each marked with a number, so that every selection of counters from it—drawing one or more counters and adding up the numbers on them—yields a different sum. The sum of the numbers on all the counters is the Total. The largest number on a counter is the Max.

'1, 2, 3' is not a Mixed Bag, because $1 + 2 = 3$. '3, 8, 72' is a Mixed Bag, whose Total is 83 and whose Max is 72.

I asked Nik, Rik, and Dik each to make up a Mixed Bag for the same number n. Nik's bag was to have the smallest possible Max and, subject to that, the largest possible Total. Rik's was also to have the smallest Max and, subject to that, the smallest Total. Dik's was to have the smallest Total and, subject to that, the smallest Max. All three bags were different.

How small could n have been?

K5   Nice price-lists

My shop sells three items only—Cheese-straws, Curry-puffs and Coffee-fudge—in packets for which I charge 4p, 6p, and 7p respectively. I call this a 'nice' price-list, because if a customer puts down 17p on the counter it *must* be one packet of each that he wants. ($2 + 3 + 6 = 11$, for instance, isn't nice, because 11 is also the sum of $3 + 3 + 3 + 2$, etc.). And 17, the total of my list, is the smallest possible for a nice list for 3 items.

What is the smallest nice price-list you can suggest for me next week, when I plan to sell packets of Kipper-paté and Tomato-sausage as well?

K6   Very nice price-lists

I later gave up selling Kipper-paté (see previous problem), and now have only 4 items on sale, but I have got dissatisfied with a merely nice price-list. Its drawback—taking my old $4 + 6 + 7$ for 3 items as an example—is that, though 17p is unambiguous, other amounts are not: 12p, for instance, could be meant for 3 packets of Cheese-straws or 2 of Curry-puffs. So I want to move to a 'very nice' price-list, where every number of pence up to the total price of one of each of the 4 items is unambiguous—it must represent a unique combination of packets (unless of course it does not represent any possible combination).

What is the smallest 'very nice' list you can recommend?

K7   Disordered numbers

Stanley's children are all twins—boy-and-girl pairs of various ages. One Christmas he gave each child a set of counters marked 1, 2, 3, ... up to n (n being the child's age in years) and asked each to arrange his counters in as 'disordered' a way as possible. Boys were to put their counters in a line, girls to put theirs in a ring. The object was to have as few numbers as possible in their natural order, going along (or once round) in either direction. You can see in Fig. 115 two early

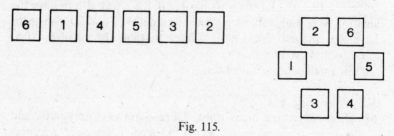

Fig. 115.

attempts by the 6-year-olds: Martin has got down to 4 (2346 or 2356) and Myfanwy to 5 (13456 or 23456). Both did better later.

Andy and Annie, the teenage twins, produced perfect results: each had the same maximum of numbers in order—Andy's in line and Annie's in a ring, of course. How old are they?

K8 Non-dividing lists

Out of the whole numbers from 1 to 50 inclusive, make a list of as many as possible, in such a way that no number on your list exactly divides any other. You could not, therefore, have 8 and 16 on your list, for instance: but you could have 8 and 12, or 12 and 16.

(1) How many numbers can you get on your list?

(2) You will find more than one way of compiling the longest possible list. What is the list that adds up to the smallest total? And what does it add up to?

(3) Consider the same questions for Non-Dividing Lists out of the whole numbers up to $2n$.

K9 Egyptian fractions

With the solitary exception of $\frac{2}{3}$, the ancient Egyptians are said to have used only fractions with a numerator of 1. Such 'unitary' fractions are very flexible tools. For instance, every unitary fraction $1/k$ is the sum of two unitary fractions in more than one way.

Represent the following as sums of two unitary fractions $(1/a + 1/b)$ in as many ways as possible:

(1) $\frac{1}{30}$

(2) $\frac{1}{31}$

(3) $\frac{2}{31}$

(4) $\frac{3}{31}$

(5) $\frac{4}{31}$

K10   Hard egyptian fractions

Most fractions that are not the sum of two unitary fractions are the sum of 3: in other words, even if

$$\frac{n}{k} = \frac{1}{a} + \frac{1}{b}$$

is not possible,

$$\frac{n}{k} = \frac{1}{a} + \frac{1}{b} + \frac{1}{c}$$

is possible. At least, that is true for small values of n.

What is the *smallest* fraction n/k for which

$$\frac{n}{k} = \frac{1}{a} + \frac{1}{b} + \frac{1}{c}$$

is not possible?

('Smallness' here is measured by the sum of numerator and denominator, $(n+k)$. $\frac{9}{19}$ is 'smaller' than $\frac{4}{25}$, which is smaller than $\frac{17}{18}$).

K11   Easy evaluation

Evaluate $(14 + \underline{/4})^{\sqrt{4}}$. The answer is 1444. How many such equalities can you find, using only. 2 or 3 digits? The right answer must be produced by simply writing out the same digits in the same order.

K12   Square roots made easy

The square root of 60,481,729? Split the number into two halves and add—

$$\begin{array}{r} 6048 \\ 1729 \\ \hline 7777 \end{array}$$ which is the right answer.

Similarly for 4,941,729. Since it has an odd number of digits, treat it as 04,941,729, and again you get the right answer 2223.

What other numbers (of up to 8 digits) yield their square roots in this painless way?

K13   Average numbers

The average of the digits in 2·250 is in fact 2·250, the same number. The same is true of $2\frac{5}{20}$. Using no more than 4 digits, can you find any other fractions or decimals with the same property?

K14   Durable numbers

Take a number: multiply its digits, giving a new number. Multiply *its* digits. And so on. You end up sooner or later with a single-digit number (perhaps 0). Some larger numbers aren't very 'durable' in this sense. 123456, for instance, gives 720 which gives 0. Let us say it has a 'durability' of 2, because it reduces to a single digit after 2 steps.

What is the smallest number with a durability of 3? Of 4? Of 5?

K15   Switch—multiplying by 5

$$\begin{array}{r} 142857 \\ 5 \\ \hline 714285 \end{array}$$

That multiplication sum is simply done by switching the last digit (7) to the beginning. If I asked you for a 12-digit number which can similarly be 'switch-multiplied by 5', you would naturally say '142857142857'; and if I asked for a 42-digit number, you could write 142857 seven times end to end. Is that the *smallest* 42-digit number that can be switch-multiplied by 5?

K   Solutions

K1S   Two stamps

System O cannot be beaten, I believe, for values of n up to 6. Thereafter, neither of Fran's systems gives the best possible set of stamp-values.

Her System O gives a maximum parcel-size of $(4n-4)$, while System T gives $n((n+6)/4)$ ($+\frac{1}{4}$ if n is odd). The following table shows the best results I have found.

n	Top Parcel Size			Stamp Set
	System O	System T	Other	
1	2	2		1
2	4	4		12 or 13
3	8	7		134
4	12	10		1356
5	16	14		13578
6	20	18		13579 10 (and others)
7	24	23	26	1349 10 12 13 (and others)
8	28	28	32	1258 11 14 15 16 (and others)
9	32	34	40	1349 11 16 17 19 20
10	36	40	44	1349 11 13 18 19 21 22
11	40	47	54	1349 11 16 18 23 24 26 27
12	44	54	64	1349 11 16 21 23 28 29 31 32
13	48	62	72	1349 11 16 20 25 27 32 33 35 36
14	52	70	78	1258 11 14 17 20 23 24 25 51 52 53
15	56	79	90	1349 11 16 20 25 29 34 36 41 42 44 45
16	60	88	100	1349 11 16 20 25 30 34 39 41 46 47 49 50
17	64	98	108	
18	68	108	118	1349 11 16 20 25 29 34 39 43 48 50 55 56 58 59
19	72	119	130	
20	76	130	142	

K2S   Two-coin currencies

(1) 17 thoils

(2) 2 and 19 in Biggleswade
   3 and 10 in Wigglesworth
   4 and 7 in Giggleswick

If only a limited number of sums are not exactly payable, the numbers of thoils represented by the two coins must be prime to each other, obviously. The general formulae for two coins a, b, (co-prime) are:

Highest sum not payable is $(ab-a-b)$

Number of sums not payable is $\dfrac{(a-1)(b-1)}{2}$

Highest sum not payable in n different ways is $(abn - a - b)$
Smallest sum payable in 2 ways is ab
Smallest sum payable in n ways is $ab(n-1)$
Number of sums payable in just one way is ab [including the sum '0']
Number of sums payable in just n ways is ab

K3S   Permo-bracelets

(1) There are four 4-permo-bracelets in 14 beads:

> 0000 1111 001011
> 0000 1111 001101 ⎫   these are 'complements'
> 0000 1111 010011 ⎬   (exchange 0 for 1 throughout)
> 0000 1011110011 ⎭

and there is a nice 14-bead solution with *two* bracelets:

> 000011 and 11110100

(2) Here is a 5-permo-bracelets in 26 beads:

> 00000, 11111, 01010, 00100, 111011

K4S   Mixed bags

$n = 5$. The bags were:

>   Nik's—6, 9, 11, 12, 13 (Total 51)
>   Rik's—3, 6, 11, 12, 13 (Total 45)
>   Dik's—1, 2, 4, 8, 16 (Total 31)

With Nik's criteria, best solutions come from the following series,
I think:

n	1	2	3	4	5	6	7	8	9	10	11	12
$f(n)$	1	1	2	3	6	11	22	42	84	165	330	654

For a Mixed bag of n counters, mark them successively with the
numbers

> $f(n)$
> $f(n) + f(n-1)$
> $f(n) + f(n-1) + f(n-2)$
> . . .
> $f(n) + f(n-1) + \cdots + f(2) + f(1)$.

With Rik's criteria, one cannot I think improve on Nik's results
for even n. But for odd n, one can reduce the next-to-smallest

counter's number to one-third its size. (e.g., for $n=5$, one has 3 instead of 9).

Dik has the easiest task. His bag contains simply $1, 2, 4, \ldots 2^{n-1}$.

K5S   Nice price-lists

16p, 24p, 28p, 30p, 31p, totalling £1.29, is I think best. The sequence

$$1 = 1$$
$$5 = 3 + 2$$
$$17 = 7 + 6 + 4$$

suggests continuing

$$49 = 15 + 14 + 12 + 8$$
$$129 = 31 + 30 + 28 + 24 + 16$$
$$\ldots$$
$$1 + (n-1)2^n = (2^n - 1) + (2^n - 2) + (2^n - 4) \ldots 2^{n-1},$$

but I have no proof.

K6S   Very nice price-lists

32p, 41, 44p, 45p, total £1.62, is best I think.

My best results (I have no proof) are:

$$1 = 1$$
$$5 = 3 + 2$$
$$26 = 10 + 9 + 7$$
$$162 = 45 + 44 + 41 + 32$$
$$1308 = 284 + 283 + 279 + 263 + 199.$$

You will see that the price-differences follow a clear pattern:

1.

1, 2.

$1, 3, 3^2$.

$1, 4, 4^2, 4^3$.

K7S   Disordered numbers

Seventeen. In a line, k^2 (or fewer) numbers can be arranged with no more than k in order. In a ring, $(k^2 + 1)$ (or fewer) can be arranged with no more than $(k+1)$ in order. So, if the two results are the same, the number of counters must be 1 more than a square, i.e. 2, 5, 10, 17, 26.... Since they were teenagers, we choose 17.

For k^2 numbers in a *line*, there are 2 different arrangements, exemplified by

$$4\ 3\ 2\ 1\ 8\ 7\ 6\ 5\ 12\ 11\ 10\ 9\ 16\ 15\ 14\ 13$$

and

4 8 12 16 3 7 11 15 2 6 10 14 1 5 9 13.

For k^2 in a *ring*, we use the second of the line-arrangements, bent round into a circle, with the number $(k^2 + 1)$—here, 17—tucked in at the join.

For fewer numbers in line or ring, we omit as many numbers as necessary from the next higher arrangement for k^2 or $(k^2 + 1)$—making the obvious adjustments if we haven't omitted the highest ones.

Though these results seem to be best, I lack proofs.

K8S   Non-dividing lists

(1)   25 numbers.

(2)   8, 12, 14, 17, 18, 19, 20, 21, 22, 23, 25, 26, 27, 29, 30, 31, 33, 35, 37, 39, 41, 43, 45, 47, 49; totalling 711.

(3)   Out of the first $2n$ integers, you can always get a ND List of n. The numbers from $(n+1)$ to $2n$ inclusive provide the list with the largest possible total, which is $n(3n+1)/2$. If $n > 1$, a list with a smaller total is always possible.

Call the smallest-total list $S(2n)$, and its total $T(2n)$. For instance, $S(4) = 2$, 3; and $T(4) = 5$. Then $S(2n)$ includes:

(a)   for all odd primes p less than $\sqrt{2n}$, $p \times S(2d)$, where $2n = 2dp \pm 0$, $1, 2 \ldots, (p-1)$;

(b)   the same multiples of 2 as (a) gives multiples of 3;

(c)   the primes between $2n/3$ and $2n$.

Applying this to $2n = 50$, we need to put $p = 3$, 5 and 7 in turn; and calculate thus:

$50 = 16.3 + 2$: $S(16) = 4$, 6, 7, 9, 10, 11, 13, 15

$\therefore$ $S(50)$ includes $3 \times (4, 6, 7, 9, 10, 11, 13, 15)$ and $2 \times (4, 6, 7, 9, 10, 11, 13, 15)$

$50 = 10.5 + 0$: $\therefore$ $S(50)$ includes $5 \times (4, 5, 6, 7, 9)$

$50 = 8.7 - 6$: $\therefore$ $S(50)$ includes $7 \times (2, 3, 5, 7)$

If you then add the primes between $16\frac{2}{3}$ and 50, you get the list at (2). Perhaps simpler for calculation is the fact that, if $2n = 6k \pm 0$, 2 (i.e. k is the nearest integer to $n/3$), then $S(2n)$ consists of:

(a)   the odd numbers from $(2k + 1)$ to $(2n - 1)$ inclusive: and

(b)   even numbers which are double all the numbers in $S(2k)$.

As to the total of $S(2n)$, we find that

$$T(6k) = 8k^2 + 2T(2k)$$
$$T(6k - 2) = (2k - 1)(4k - 1) + 2T(2k)$$
$$T(6k + 2) = (2k + 1)(4k + 1) + 2T(2k)$$

K9S Egyptian fractions

(1) $\frac{1}{30} = \frac{1}{930} + \frac{1}{31} = \frac{1}{480} + \frac{1}{32} = \frac{1}{330} + \frac{1}{33} = \frac{1}{255} + \frac{1}{34} = \frac{1}{210} + \frac{1}{35}$
$= \frac{1}{180} + \frac{1}{36} = \frac{1}{130} + \frac{1}{39} = \frac{1}{120} + \frac{1}{40} = \frac{1}{105} + \frac{1}{42} = \frac{1}{90} + \frac{1}{45}$
$= \frac{1}{80} + \frac{1}{48} = \frac{1}{75} + \frac{1}{50} = \frac{1}{66} + \frac{1}{55} = \frac{1}{60} + \frac{1}{60}$

(2) $\frac{1}{31} = \frac{1}{992} + \frac{1}{32} = \frac{1}{62} + \frac{1}{62}$

(3) $\frac{2}{31} = \frac{1}{496} + \frac{1}{16} = \frac{1}{31} + \frac{1}{31}$

(4) No solution

(5) $\frac{4}{31} = \frac{1}{248} + \frac{1}{8}$

In general, $1/k = ((1/a) + (1/b))$ has $\frac{1}{2}(1 + (2l + 1)(2m + 1)(2n + 1) \ldots)$ solutions, when $k = 2^l \times p^m \times q^n \ldots$ (p, q, ... being distinct primes). Hence the 14 solutions in (1), and the 2 solutions for prime k (as in (2)).

$n/k = ((1/a) + (1/b))$ is soluble if and only if k^2 has a factor $X \leqslant k$ such that n will divide $(k + X)$ exactly. For each such X, $a = (k + X)/n$ and $b = [k + (k^2/X)]/n$.

If the denominator k is *prime*, $n/k = ((1/a) + (1/b))$ is soluble if and only if n divides $(k + 1)$ exactly.

K10S Hard egyptian fractions

$\frac{8}{11}$ is the smallest I have found. $\frac{8}{17}$ and $\frac{9}{19}$ are other small examples. If $n = 7$ or less, n/k seems always to be the sum of 3 unitary fractions, at least for k up to 300.

K11S Easy evaluation

The ones I have found are:
$$24 = \underline{/2\sqrt{4}} = \underline{/2 + \sqrt{4}} = \underline{/\sqrt{2^4}} = \underline{/2^{\sqrt{4}}}$$
$$36 = \underline{/3 \times 6}$$
$$120 = \underline{/1 \div \cdot 2} \pm 0 \text{ (and similarly for 121, 122, ... 129)}$$
$$127 = -1 + 2^7$$
$$144 = \underline{/-1 + 4 \times /4}$$
$$216 = \sqrt{(\underline{/2 + 1})^6} = (\cdot 2 + \cdot 1) \times \underline{/6}$$
$$288 = 2^8 \div \cdot 8$$
$$343 = (3 + 4)^3$$
$$355 = 3\underline{/5} - 5 = \underline{//3} \times \cdot 5 - 5$$

$$360 = \underline{/3} \times 60$$
$$660 = \underline{/6} - 60$$
$$693 = \underline{/6} - (9 \times 3)$$
$$720 = \underline{/7 - 2}^{\circ}$$
$$729 = (7 + 2)^{\sqrt{9}}$$
$$744 = (7 + \underline{/4}) \times \underline{/4}$$

K12S Square roots made easy

$$81 \to 9 \qquad\qquad 1 \to 1$$
$$9{,}801 \to 99$$
$$998{,}001 \to 999$$
$$99{,}980{,}001 \to 9{,}999$$

$3{,}025 \to 55$	$2{,}025 \to 45$
$494{,}209 \to 703$	$88{,}209 \to 297$
$52{,}881{,}984 \to 7{,}272$	$7{,}441{,}984 \to 2{,}728$
$25{,}502{,}500 \to 5{,}050$	$24{,}502{,}500 \to 4{,}950$

Note that, like those quoted in the question, they come in complementary pairs.

K13S Average numbers

I think the following list is practically complete: $\quad 1 = \frac{1}{1} = \frac{11}{11} = \frac{20}{20}$

$$1\tfrac{1}{2} = 1\cdot500 \qquad 2 = \tfrac{10}{5} \qquad 2\tfrac{1}{4} = 2\tfrac{5}{20} = 2\cdot250 \qquad 3 = \tfrac{36}{12} = \tfrac{63}{21} = \tfrac{90}{30} \qquad 3\tfrac{1}{2} = \tfrac{65}{20}$$
$$3\tfrac{3}{4} = \tfrac{90}{24} = 3\tfrac{9}{12} = 3\cdot750 \qquad 4 = \tfrac{24}{6} = \tfrac{56}{14} = \tfrac{92}{23} \qquad 4\tfrac{1}{4} = 4\tfrac{8}{32} \qquad 4\tfrac{1}{2} = \tfrac{81}{18} = 4\cdot5 \qquad 4\tfrac{2}{3} = 4\tfrac{4}{6}$$
$$5 = \tfrac{35}{7} \qquad 5\tfrac{2}{3} = 5\cdot66 \qquad 6 = \tfrac{54}{9} \qquad 8 = 7\cdot\dot9.$$

K14S Durable numbers

Durability 3: $39 - 27 - 14 - 4$.

Durability 4: $77 - 49 - 36 - 18 - 8$.

Durability 5: $688 - 384 - 96 - 54 - 20 - 0$ is not quite the smallest;
$$679 - 378 - 168 - 48 - 32 - 6 \text{ is.}$$

K15S Switch-multiplying by 5

No: 102040, 816326, 530612, 244897, 959183, 673469, 387755 is.

If the number required is abcd ... vwx, let us say that abcd ... vw = A, containing n digits. We must solve

$$5(10A + x) = 10^n x + A,$$
$$\text{i.e.} \quad A = x\,\frac{(10^n - 5)}{49}.$$

Now if x (the 'switched' digit) $= 7$, $A = (10^n - 5)/7$, and we divide 7 into 1000 . . . till we get a remainder 5, thus:

$$7 \,|\, \overline{100000}$$
$$14285 + 5.$$

The dividend was 10^5, and there are 5 digits in 14285, so $n = 5$ and 14285 (7) is an answer.

But we can also divide 49 into 1000 . . ., to see if that will yield other answers. It does. $(10^{41} - 5)/49 = 2040$, 816326, 530612, 244897, 959183, 673469, 387755. But that has only 40 digits. To make A have the necessary 41 digits, we must have $x = 5$, 6, 7,8 or 9. $x = 5$ produces the smallest answer, given above. $x = 6$, 8 and 9 produce other answers, with the same digits cyclically permuted. $x = 7$ produces our old friend, 142857 repeated.

L More Various

L1 Odd man out
Four of the five pictures in Fig. 116 have a common characteristic, which the fifth has not. Which is the odd man out?

L2 Ordering weights
Suppose you have 3 objects, A, B and C; each has a different weight; and you want to arrange them in order of weight. Your method is to weigh them in pairs on a balance. How many weighings will you need?

As economical a way of proceeding as you can find can be set out as follows:

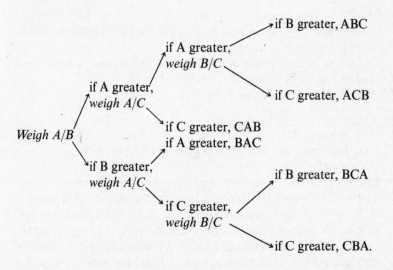

Weigh A/B

if A greater, weigh A/C
    if A greater, weigh B/C
        if B greater, ABC
        if C greater, ACB
    if C greater, CAB

if B greater, weigh A/C
    if A greater, BAC
    if C greater, weigh B/C
        if B greater, BCA
        if C greater, CBA.

The final 'ABC' etc. indicates the determined order of the objects,

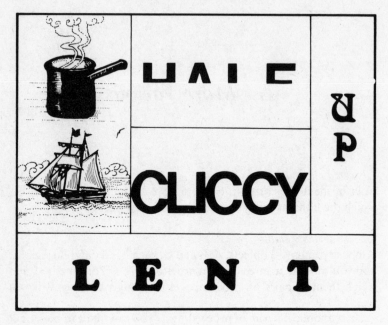

Fig. 116.

heaviest first. As you see, you may succeed with 2 weighings. But at worst you need 3.

If, given the best procedure to order n objects, you may at worst need w weighings, we can say $w = W(n)$.

$W(1) = 0$, clearly: $W(2) = 1$: $W(3)$ is 3.

(1) What is $W(6)$?
(2) What is $W(10)$?
(3) How near can you get to a general formula for $W(n)$?

L3 Darts scores
(1) I watched Hermann, Ian and John start a three-handed game of darts, 1000 up. Each got quickly off the mark, but I got bored and went to the bar for a pint. I looked across briefly 5 times during the game, and on each occasion I found a curious thing—the total of their scores at the time was a square number, and so was the sum of any two of their scores. The total of Hermann's and Ian's on one

occasion was the same as the total of Hermann's and John's on a later occasion. What were the scores each time I looked?

(2) If there are *four* players, and the sum of the scores of any two is a square number (though the total of all four isn't necessarily a square), what is the smallest possible score-total?

L4 π^e and e^π

(1) Which is bigger, π^e or e^π?

(2) Which is bigger, $(3)^{2\frac{1}{2}}$ or $(2\frac{1}{2})^3$?

(3) More generally, which is greater, A^b or b^A, where A is greater than b and both A and b are positive (and may or may not be integral)?

(4) When, if ever, does $A^b = b^A$ (apart from the times when $A = b$, of course)?

L5 *Various brackets*

We can insert brackets in the expression $1 \div 2 \div 3$ so as to give it one of *two* values, viz.

either $\frac{1}{6}(=(1 \div 2) \div 3)$
  or $\frac{3}{2}(=1 \div (2 \div 3))$.

(1) How many different values can be made by inserting brackets in different ways in the expression

$$1 \div 2 \div 3 \div 4 \div 5 \div 6 \div 7 \div 8?$$

(2) And how many with

$$1 \div 2 \div 3 \ldots \div 11 \div 12?$$

L6 *How many can you win?*

The four of them sat in a circle at the *Towers of Ilium* thoughtfully sipping whiskey: Dai O'Tribe, the vituperative Celt; Mr Justice ('Dumsy Benny') Jesserint; the builder, 'Storied' Ern; and Annie May ('Ted') Bust, the filmstar.

'You can't win 'em all', said Dai.

'You can't win any of 'em', said Benny.

'Indeed you can't win any of 'em', said Ern.

'An odd number of us', said Ted 'are not telling the truth'.

What conclusion can you draw from this conservation about how many of 'em it is possible to win?

L7   *Dot-wiggling*

(1) Fig. 117 shows paths joining:
   19 dots in a 3-hexagon (3 dots along each edge).
   27 dots in a 3/4-hexagon (3 and 4 dots along alternate edges)
   28 dots in a 7-triangle (7 dots along each edge)

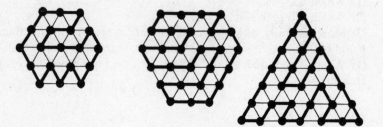

Fig. 117.

I ask you to join all the dots in each diagram to form a single minimum-length reentrant path which is as wiggly as possible. The paths in the figure are clearly minimum-length, 19, 27 and 28 units long respectively. As to wiggles, the paths shown have 14, 17 and 15 bends. Can you get nearer to the optimum of 19, 27 and 28 bends?

(2) How well can you do with other *n*-hexagons, $n/(n+1)$-hexagons, and *n*-triangles?

L8   *Truthful friends*

At last night's meeting of the Truth-Tellers' Club, no outsiders were present. I over-heard the following remarks:

NIK: Every pair of people here are either friends of each other or not.
OZ: Every pair of friends here have no common friends here.
PAM: Every pair here who aren't friends have exactly two common friends here.
ROJ : There are fewer than 22 people here.
SI: I wish I could think of an intelligent remark to make.
All these remarks were true, of course.

How many people were at the meeting? Can you tell how many
friends of Si were among them?

L   Solutions

L1S   Odd man out
Up. Each of the other four is a word or phrase treated in the way
suggested by the word or phrase itself; they are

> Canon Spooner (Pan and Schooner)
> HALF
> CYCLIC
> ALTERNATE.

Up, on the other hand, is written 'down', not 'up'.

L2S   Ordering weights
(1) W(6) is 10
(2) W(10) is 23
(3) It is quite easy to set narrow limits for $W(n)$, as follows. Consider
$n = 7$, say. The number of different possible weight-orders is 7!, or
5040. With 12 weighings, each of which can have 2 results, we can
distinguish between at most 2^{12} or 4096 orders. So clearly W(7) is
greater than 12. So if $K(n) = k$, where 2^k is the smallest power of 2
which is equal to or greater than $n!$, $K(n)$ is the minimum possible
value for $W(n)$.

And we can also set a maximum value. If we have already an
ordered set of 7 (or fewer), we can clearly mesh one new object with
that set in 3 weighings: we weigh it first against the middle one of the
7, then against the 2nd or 6th, and finally against the 1st/3rd/5th/7th.
So if l is the smallest power of 2 which is equal to or greater than
$(n-1)$,

$W(n) \leqslant W(n-1) + l$. This gives a series of maximum values for
$W(n)$. Let us call this series $L(n)$.

The following table lists these values.

n	$K(n)$	$L(n)$	$M(n)$	$N(n)$	$W(n)$
1	0	0			0
2	1	1			1
3	3	3			3
4	5	5			5
5	7	8	7		7
6	10	11	10		10
7	13	14	13		13
8	16	17	16		16
9	19	21	20		20?
10	22	25	24	23	23?
11	26	29	28	27	27?
12	29	33	32	31	31?
13	33	37	36	35	35?
14	37	41	40	39	39?

For n up to 4, $L(n) = K(n)$, so no improvement is possible. For $n = 5$, we can do better, as follows:

    Weight A/B: say A is greater;

    Weigh C/D: say C is greater;

    Weigh A/C: say A is greater.

So we know (bigger weights on left)

Then weigh C/E, and

    (a) if E greater, weigh B/C and E/A and (if necessary) B/D or B/E, or

    (b) if C greater, weigh D/E and B/D and either B/C or B/E.

This orders 5 in 7 weighings. Since $K(5) = 7$, $W(5) = 7$. And, considering how $L(n)$ was constructed, we can use this result to lower its values by 1 from $n = 5$ onwards. The results are listed as $M(n)$ in the table. This achieves $K(n)$ for n up to 8.

For $n = 9$, I cannot improve on $M(9) = 20$, though $K(9) = 19$.

For $n = 10$, though I cannot achieve $K(10) = 22$, a result in 23 (one better than $M(10) = 24$) is possible, as follows. First order 8 objects

and 2 separately. We have used $16 + 1 = 17$ weighings, and know that
A – B – C – D – E – F – G – H
X – Y

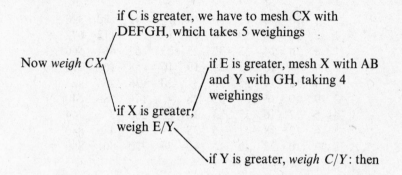

if C is greater, we have to mesh CX with DEFGH, which takes 5 weighings

Now *weigh CX*

if E is greater, mesh X with AB and Y with GH, taking 4 weighings

if X is greater, weigh E/Y

if Y is greater, *weigh C/Y*: then

$\begin{cases} \text{if C greater, mesh D with Y and X with AB, taking 3 weighings} \\ \text{if Y is greater, mesh XY with AB, taking 3 weighings.} \end{cases}$

Hence $N(n) = M(n) - 1$, from $n = 10$ onwards.

I have not been able to make further improvements, though I have seen W(12) quoted, without giving details, as 30, whereas $N(12) = 31$.

L3S   Darts scores
(1)

41J	80H	320I $(80 + 320 = 400)$
57J	112H	672I
136	264	825I $(136 + 264 = 400)$
385	456	840I
720	801	880I

(2) $8840 = 5378 + 2018 + 1346 + 98.$

(1)   If $\begin{cases} a+b+c = D^2 \\ a+b = E^2 \\ b+c = F^2, \\ a+c = G^2 \end{cases}$ then $\begin{cases} a = D^2 - F^2 \\ b = D^2 - G^2 \\ c = D^2 - E^2 \end{cases}$ and $\begin{aligned} 2D^2 &= E^2 \\ &+ F^2 + G^2. \end{aligned}$

So if $D = E + x = F + y = G + z$,
$2D^2 = (D-x)^2 + (D-y)^2 + (D-z)^2$, and
$D = x + y + z + Q$
$Q^2 = 2xy + 2(x+y)z.$

Assuming $x < y < z$, trial produces the following possible sets of scores:

x	y	z	Q	D	E	F	G	a	b	c
1	2	10	8	21	20	19	11	41	80	320
1	2	16	10	29	28	27	13	57	112	672
1	8	10	14	33	32	25	23	65	464	560
2	4	7	10	23	21	19	16	88	168	273
2	4	15	14	35	33	31	20	136	264	825
2	9	10	16	37	35	28	27	144	585	640
4	5	12	16	37	33	32	25	280	345	744
5	6	12	18	41	36	35	29	385	456	840
8	9	10	22	49	41	40	39	720	801	880

and only one series of five satisfies the conditions.

(2) If $a+b+c+d = M = Q^2+R^2 = S^2+T^2 = V^2+U^2$, $a+b = Q^2$, $c+d = R^2$, $a+c = S^2$, $b+d = T^2$, $b+c = V^2$, $a+d = U^2$;

$$a = \frac{Q^2+S^2-V^2}{2}, \qquad b = \frac{Q^2-S^2+V^2}{2}, \qquad c = \frac{-Q^2+S^2+V^2}{2},$$

$$d = \frac{R^2-S^2+U^2}{2} = \frac{2M-(Q^2+S^2+V^2)}{2}.$$

Trial shows the difficulty of making a, b, c and d all positive. But we do not have to go very far before finding the answer. (Actually, the problem is easier if we require M to be square. For the smallest square number which is the sum of two squares in three or more ways is $5^2 \times 13^2 = 4225$. This yields the answer $2607\frac{1}{2} + 992\frac{1}{2} + 528\frac{1}{2} + 96\frac{1}{2}$, and we have only to multiply by 4 to get $16,900 = 10,430 + 3,970 + 2,114 + 386$.)

L4S π^e and e^π

(1) e^π is bigger.

(2) $(2\frac{1}{2})^3$ is bigger. It is the square root of $15625/64 = 244 \cdot 14$ while $(3)^{2\frac{1}{2}}$ is the square root of 243.

(3) The graph in Fig. 118 gives the general picture.

Both the lines in the graph show values for which $A^b = b^A$: along

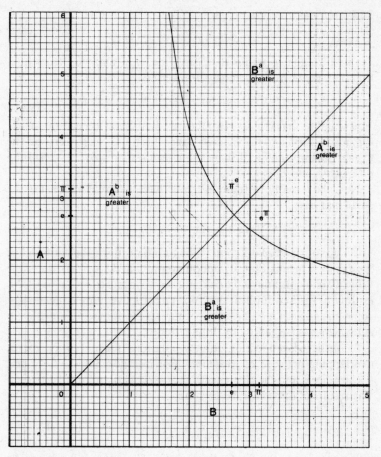

Fig. 118.

the curved line, A is not equal to b: the lines cross at $A=b=e$ $(=2\cdot718\ldots)$

(4) $A=4$, $b=2$ gives equality: so do $A=3\sqrt{3}$, $b=\sqrt{3}$. These and an infinity of other solutions can be derived from $b=c^{1/(c-1)}$, $A=c^{c/(c-1)}$ $=cb$; or (what comes to the same thing)

$$b=\left(1+\frac{1}{m}\right)^{m}; \quad A=\left(1+\frac{1}{m}\right)^{m+1}.$$

L5S Various brackets
(1) 60
(2) 216

We need to consider this in stages. First, how many different fractions can result from variously bracketing

$$a \div b \div c \div d \ldots ?$$

You will find that the first two terms, a and b, always end up on the top and bottom line respectively: the others, c, d etc, can end on the top or the bottom. So, if there are n terms, there are 2^{n-2} resulting possible fractions.

Next, how many of these fractions have different values for the particular case

$$1 \div 2 \div 3 \div 4 \ldots \div n ?$$

For n up to 7, all are different. When $n = 8$ or more, the equivalence $3 \times 8 = 4 \times 6$ means that a fraction containing 3.8/4.6 has the same value as one containing 4.6/3.8. For $n = 9$ or more, $3 \times 4 \times 6 = 8 \times 9$, leading to further duplications. With $n = 10$, five more equivalences come in, and with $n = 12$ we get 11 more. If the number of terms in an equivalence is k, the loss by duplication is 2^{n-k-2}.

A further complication affects solutions for $n = 14$ or more; overlapping between equivalences. For instance, $3 \times 14 = 6 \times 7$ and $4 \times 10 = 5 \times 8$, and the elements of the two equations are all different. The results of these calculations are

n	2	3	4	5	6	7	8	9	10	11	12
Number of different values	1	2	4	8	16	32	60	116	180	360	216

L6S How many can you win?
No conclusion at all, I think. But many problem-composers would say you can conclude that it is possible to win them all. They would argue something like this. The conclusion (if any) must be one of 3 propositions, one of which must be true:

(A) 'You can win 'em all'; in which case D, B and E are lying, and T may or may not be lying:

(B) 'You cannot win any of 'em'; in which case D, B and E are telling the truth—but what about T? If she lies, there is one lie, so her statement is true. If she tells truth, there are no lies, so her statement is false. So her statement is neither true nor false.

(C) 'You can win some of 'em, but not all'; in which case D speaks true, B and E lie—but, again, what about T? Just as in case (B), her statement is neither true nor false.

Therefore (A) must be true, the argument runs, since otherwise T's statement is neither true nor false but meaningless.

I find this unconvincing. If, as here, we find that all quoted statements cannot be true, I think a more natural assumption than that all statements are meaningful (i.e. either true or false) would be that as many statements as possible are true. Which would lead to the conclusion that you can't win any of 'em—and that what Ted said was neither true nor false but nonsense.

The easiest way, anyway, out of this kind of mess would be for problem-putters who want solvers to work on some assumption about the truth of quoted statements (other than the assumption that all are true) to state the assumption that they are to work on.

L7S   Dot-wiggling

Fig. 119 shows the wiggliest answers I can find, with 17, 26 and 25 bends respectively.

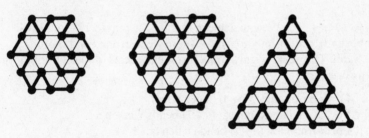

Fig. 119.

An n-hexagon contains $1 + 3n(n-1)$ dots, and I believe a path with the full $1 + 3n(n-1)$ bends is possible for all n except 3, where we have to settle for one fewer.

An $n/(n+1)$-hexagon contains $3n^2$ dots. Again, I believe a path with $3n^2$ bends is possible except for $n = 3$.

An n-triangle contains $[n(n+1)]/2$ dots. When n exceeds 2 one bend must be lost at each corner, so we can never improve on

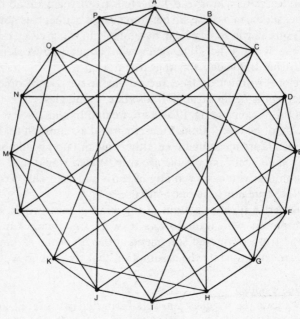

Fig. 120.

$(n-2)(n+3)/2$. We can achieve that for all n up to 12 (and, I believe, beyond), except only for $n=4$ where we lose one bend more.

Note: The least-wiggly-route problem seems duller. I think the answers are:

> For n-hexagons, $(5n-3)$ bends when $n>1$
> For $n/(n+1)$-hexagons, $(4n+1)$ bends when $n>1$
> For n-triangles, $(n+2)$ bends when $n>3$.

L8S   Truthful friends

16: and Si (like the rest) had 5 friends present. Suppose there are n people present in all. Suppose Si has m friends present. By Oz's rule, none of those m are friends of each other. So, by Pam's rule, every pair of them have 2 common friends: one is Si, so each pair of m has just one friend among the 'rest' (who number $n-m-1$). Can two of those pairs have the same friend among the rest? If they did, that friend and Si would have 3 or 4 common friends. So each pair has a different common friend among the rest. So the number of the rest $(n-m-1)$ is at least as big as the number of pairs among m (i.e., $[m(m-1)]/2$).

Now apply Pam's rule to Si and each member R of the rest. Each Si/R pair must have two common friends, and they must obviously be among the m who are friends of Si. Can two members R share with Si the same pair of common friends? No, because then that pair would have *more* than 2 common friends (Si and 2 R's). So each member of R shares with Si a different pair of common friends among m. So $[m(m-1)]/2$ is at least as big as $(n-m-1)$.

Hence $[m(m-1)]/2 = n-m-1$; i.e., $n = 1 + [m(m+1)]/2$.

Since the same argument applies to others present just as well as to Si, it follows that each person present has the same number of friends, and that possible values for n and m must come from the following table:

n (number present)	2	4	7	11	16	22	29	37...
m (number of friends each)	1	2	3	4	5	6	7	8...

The data require a solution with $5 \leqslant n < 22$. $n = 7$ does not yield a solution, for $7 \times 3/2$ 'friendships' is impossible. And it is not hard to show that $n = 11$ does not yield a solution either. But $n = 16$ does, as Fig. 120 (where the lines represent friendships) shows.